G000271430

AutoC/ Reference
6th Edition

by
Cheryl R. Shrock
Professor, retired
Drafting Technology
Orange Coast College, Costa Mesa, Ca.
Autodesk Authorized Author

Updated for AutoCAD 2013 and 2014
by
Steve Heather
Former Lecturer of
Mechanical Engineering &
Computer Aided Design

INDUSTRIAL PRESS INC.

Copyright ©2014 by Cheryl R. Shrock and Industrial Press

ISBN 978-0-8311-3484-6

Printed in the United States of America
All rights reserved. No part of this book may be reproduced or
transmitted in any form or by any means, electronic or
mechanical, including photocopying, recording, or by any
information storage and retrieval system, without written
permission from the publisher.

Limits of Liability and disclaimer of Warranty
The author and publisher make no warranty of any kind,
expressed or implied, with regard to the documentation contained
in this book.

Autodesk, Autodesk 360, AutoCAD, Design Web Format, and
DWF are either registered trademarks or trademarks of
Autodesk, Inc., in the U.S.A. and / or certain other countries.
Certain content provided courtesy of Autodesk, Inc., © 2013.
All rights reserved.

Industrial Press Inc.
32 Haviland Street, Unit 2C
South Norwalk, CT 06854

10 9 8 7 6 5 4 3 2 1

Why do you need this book?

Refresh your memory or learn something new.
No need to memorize.
Handy size and easy to use.

*"I originally wrote this book for myself. Occasionally I forget commands and
"How to" steps also. Its convenient small size allows me to toss it in my
briefcase or set it beside my monitor. No need to memorize anymore."*

Cheryl Shrock

About this book

The *AutoCAD Pocket Reference, 6th Edition*, includes all the important
fundamental Commands, Concepts, and How to information for the every
day use of AutoCAD. It is not designed to take the place of larger textbooks
but rather to supplement them as a quick reference.

**Note: If you are using AutoCAD 2012 or 2011, please refer to the
Fifth Edition of the AutoCAD Pocket Reference.**

How to use this book

The information in this book has been organized in 13 sections.
Each section contains related material. For example, if you needed
information regarding dimensioning, you would go to:

Section 3: Dimensioning.

A comprehensive Table of Contents and a cross-referenced Index have
been carefully prepared to insure easy access to all information.

About the Authors

Cheryl R. Shrock, formerly Professor and Chairperson of Computer Aided
Design at Orange Coast College in Costa Mesa, California. She is also an
Autodesk® registered author. Cheryl is always trying to think of new ways to
make it easy to learn AutoCAD. The Pocket Reference series is the latest in
that endeavor.

Steve Heather has 30-plus years of experience as a practicing Mechanical
Engineer and has taught AutoCAD to Engineering and Architectural
students at the college level. He is an authorized AutoCAD beta tester.
Steve lives near Canterbury, England and welcomes taking your questions
or comments at: steve.heather@live.com

AutoCAD Books by Cheryl R. Shrock:

Currently Available*

Beginning AutoCAD **2011** .... ISBN 978-0-8311-3416-7
Advanced AutoCAD **2011** .... ISBN 978-0-8311-3417-4

Beginning AutoCAD **2012** .... ISBN 978-0-8311-3430-3
Advanced AutoCAD **2012** .... ISBN 978-0-8311-3431-0

Beginning AutoCAD **2013** .... ISBN 978-0-8311-3456-3
Advanced AutoCAD **2013** .... ISBN 978-0-8311-3457-0

Beginning AutoCAD **2014** .... ISBN 978-0-8311-3473-0
Advanced AutoCAD **2014** .... ISBN 978-0-8311-3474-7

AutoCAD Pocket Reference
5th Edition, Releases 2011/2012......... ISBN 978-0-8311-3428-0

AutoCAD Pocket Reference
6th Edition, Releases 2013/2014......... ISBN 978-0-8311-3484-6

* As of Fall, 2013

For information about these books visit: www.industrialpress.com

TABLE OF CONTENTS

The following Table of Contents is an overview of the contents within this book. Refer to the Index at the back of this book to easily locate specific information.

SECTION 1 - Action Commands

Array	1-2
Break	1-12
Chamfer	1-14
Copy	1-16
Divide	1-20
Erase	1-21
Explode	1-22
Extend	1-23
Fillet	1-24
Match Properties	1-26
Measure	1-27
Measuring Tools	1-28
Mirror	1-29
Move	1-31
Nudge	1-33
Offset	1-34
Rotate	1-36
Scale	1-37
Stretch	1-38
Trim	1-39
Undo/Redo	1-40
Wipeout	1-41
Zoom	1-42

SECTION 2 - Concepts

Model and Layout options	2-2
Model and Layout tabs	2-3
Why Layouts are useful	2-5
Creating scaled drawings	2-6
Adjusting the viewport scale	2-7

SECTION 3 - Dimensioning

Dimensioning 3-2
Dimension Styles 3-4
Editing Dimensions 3-14
Linear, Baseline, Continue.................... 3-20
Aligned and Angular.............................. 3-23
Arc's 3-25
Diameter and Radii 3-28
Flip Arrow.. 3-32
Quick Dimensions 3-33
Breaks and Jogs 3-35
Multileader 3-40
Ignoring Hatch Objects 3-44
Ordinate Dimensioning 3-45

SECTION 4 - Drawing Entities

Arc.. 4-2
Blocks... 4-9
Multileader and Blocks........................... 4-17
Centermark 4-22
Circle 4-23
Donut.. 4-25
Ellipse... 4-26
Hatch.. 4-29
Gradient 4-38
Lines... 4-40
Point... 4-43
Polygon 4-44
Polylines... 4-45
Rectangle.. 4-51
Revision Cloud....................................... 4-54

SECTION 5 - How to.....

Add a Printer / Plotter........................... 5-2
Create a Page Setup 5-6
Create a Viewport 5-9
Use a Layout... 5-13
Create a Template 5-15
Use a Template...................................... 5-17

Open an existing File 5-18
Open Multiple Files 5-19
Save a Drawing 5-22
Start a New Drawing 5-25
Create an Autodesk Account 5-26
Save a file to Autodesk 360 Cloud 5-28
Exit AutoCAD 5-30
Customizing a Wheel Mouse 5-31
Selecting Objects 5-32

SECTION 6 - Layers

Understanding Layers 6-2
Selecting Layers 6-2
Controlling Layers 6-3
Layer Color 6-5
Lineweights 6-6
Creating New Layers 6-8
Load Linetypes 6-9
Layer Match 6-11
Layer Transparency 6-12

SECTION 7 - Input Options

Coordinate Input 7-2
Direct Distance Entry 7-6
Polar Coordinate 7-7
Polar Tracking 7-9
Polar Snap 7-10
Dynamic Input 7-11

SECTION 8 - Miscellaneous

Background Mask 8-2
Back Up Files 8-3
Grips 8-4
Object Snap 8-6
Pan 8-11
Properties Palette 8-13
Quick Properties Panel 8-15
Command Line Enhancements 8-17
AutoCorrect 8-17

SECTION 9 - Plotting

Background Plotting.............................. 9-2
Plotting from Model Space.................... 9-3
Plotting from Paper Space.................... 9-7

SECTION 10 - Settings

Drawing Setup 10-2
Drawing Limits 10-2
Units/Precision..................................... 10-5
Annotative Property 10-6
Annotative Objects............................... 10-7
Multiple Annotative Scales................... 10-12
Remove Annotative Scales................... 10-16
Annotative Hatch.................................. 10-18

SECTION 11 - Text

Creating a New Text Style 11-2
Select a Text Style 11-4
Delete a Text Style............................... 11-5
Changing a Text Style........................... 11-6
Multiline Text.. 11-7
MTJIGSTRING System Variable 11-8
Tabs, Indents and Spell Checker.......... 11-9
Columns.. 11-10
Multiline Text Line Spacing.................. 11-12
Editing Multiline Text............................ 11-13
Single Line Text 11-14
Editing Single Line Text 11-17
Special Text Characters 11-18

SECTION 12 - UCS

Displaying the UCS Icon....................... 12-2
Moving the Origin................................. 12-3
Return the Origin to WCS 12-3
Examples of Moving the Origin............. 12-4

SECTION 13 - Constraints

Parametric Drawing 13-2
Geometric Constraints 13-3
Coincident Constraint 13-5
Collinear Constraint 13-6
Concentric Constraint 13-7
Horizontal Constraint 13-10
Vertical Constraint 13-11
Controlling Icon Display, Geometric 13-15
Dimensional Constraints 13-16
Parameter Manager 13-20
User Defined Parameters 13-23
Convert Dimensional Constraints 13-24
Controlling Icon Display, Dimensional 13-25

INDEX

Section 1
Action Commands

ARRAY

The ARRAY command allows you to make multiple copies in a **RECTANGULAR** or Circular **(POLAR)** pattern and even on a **PATH**. The maximum limit of copies per array is 100,000. This limit can be changed but should accommodate most users. (Refer to the Help menu if you choose to change the limit)

RECTANGULAR ARRAY

This method allows you to make multiple copies of object(s) in a **rectangular pattern**. You specify the number of rows (horizontal), columns (vertical) and the spacing between the rows and columns. <u>The spacing will be equally spaced between copies</u>.

Spacing is sometimes tricky to understand. ***Read this carefully***. The spacing is the <u>distance from a specific location on the original to that same location on the future copy</u>. It is <u>not just the space in between</u> the two. Refer to the example below.

To use the rectangular array command you will select the object(s), specify how many rows and columns desired and the spacing for the rows and the columns.

<u>Refer to step by step instructions on page 1-3</u>.

Example of a Rectangular Array:

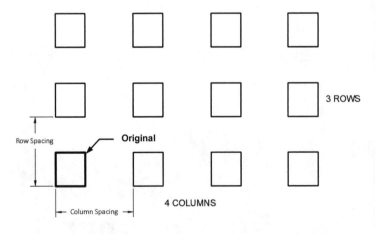

ARRAY....continued

How to create a RECTANGULAR ARRAY

1. Draw a **1"** (Inch) Square Rectangle. ⬜

2. Select the **ARRAY** command using one of the following:

Ribbon = Home tab / Modify panel / Array ▼
or
Keyboard = Array <enter>

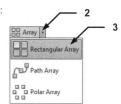

3. Select **Rectangular Array**.

4. Select Objects: *Select the Object to be Arrayed*

5. Select Objects: *Select more objects or <enter> to stop*

The **Array Creation** tab appears with a **3 x 4** grid array of the object selected.

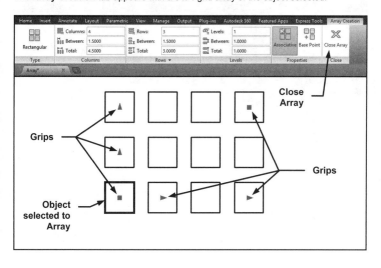

6. Make any changes necessary in the **Array Creation** tab, then press **<enter>** to display any changes.

7. If the display is correct select **Close Array**.

ARRAY....continued

How to edit a RECTANGULAR ARRAY

1. Select the Array to edit.

The **Array** panel is displayed. (The **Quick Properties** will also be displayed if you have the **QP** button **ON** in the Status bar.)

2. Make any changes necessary in the **Array Creation** tab, then press **<enter>** to display any changes.

3. If the display is correct select **Close Array**.

Array Panel

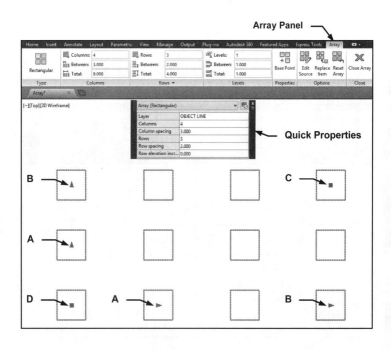

Quick Properties

Continued on the next page...

ARRAY....continued

How to edit a RECTANGULAR ARRAY

Using Grips to edit.

You may also use the Grips to edit the spacing. Just click on a grip and drag.

A. The first ▶ or ▲ allows you to change the spacing between the columns or rows.

B. The last ▶ or ▲ allows you to change the total spacing between the base point and the last ▶ or ▲ and also to add extra columns or rows, or change the axis angle.

C. The ■ allows you to change the total row and column spacing simultaneously, and also to add extra columns and rows simultaneously.

D. Use the Base Point grip ■ to **MOVE** the entire Array.

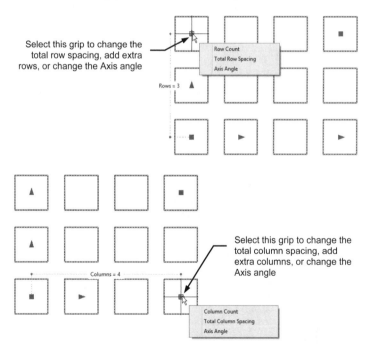

Select this grip to change the total row spacing, add extra rows, or change the Axis angle

Row Count
Total Row Spacing
Axis Angle

Rows = 3

Select this grip to change the total column spacing, add extra columns, or change the Axis angle

Columns = 4

Column Count
Total Column Spacing
Axis Angle

Continued on the next page...

ARRAY....continued

How to edit a RECTANGULAR ARRAY

Using Grips to edit.

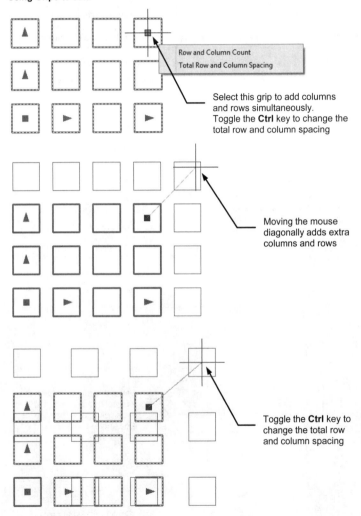

Row and Column Count
Total Row and Column Spacing

Select this grip to add columns
and rows simultaneously.
Toggle the **Ctrl** key to change the
total row and column spacing

Moving the mouse
diagonally adds extra
columns and rows

Toggle the **Ctrl** key to
change the total row
and column spacing

ARRAY....continued

POLAR ARRAY

This method allows you to make multiple copies in a circular pattern. You specify the total number of copies to fill a specific Angle or specify the angle between each copy and angle to fill.

To use the polar array command you select the object(s) to array, specify the center of the array, specify the number of copies or the angle between the copies, the angle to fill and if you would like the copies to rotate as they are copied.

Example of a Polar Array

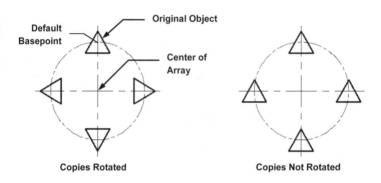

Copies Rotated   **Copies Not Rotated**

Note: the two examples shown above use the **objects default base point**.
The examples below display what happens if you specify a basepoint.

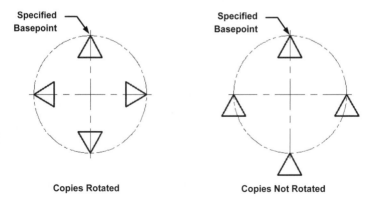

Copies Rotated   **Copies Not Rotated**

ARRAY....continued

How to create a POLAR ARRAY

Using "Number of Items".

1. Draw a 3" Radius circle.
2. Add a .50 Radius 3 sided Polygon and place as shown.
3. Select the **ARRAY** command using one of the following:

   Ribbon = Home tab / Modify panel / Array ▼
   or
   Keyboard = Array <enter>

4. Select **Polar Array**.
5. Select Objects: *Select the Object to be Arrayed. (Polygon)*
6. Select Objects: *Select more objects or <enter> to stop*
7. Specify center point of array or [Base point / Axis of Rotation] **Select the Center Point of the Circle**

   7. Snap to the center of the Circle to select the center of the Array

The **Array Creation** tab appears and the array defaults to 6 items.

8. Enter Items: **12**
9. Enter Fill: **360**
10. Press <**enter**> to display the selections
11. Select **Close Array** if display is correct

Note:

12 items were evenly distributed within 360 degrees

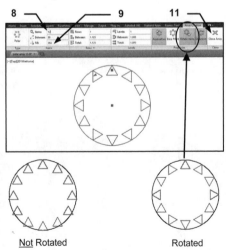

Not Rotated   Rotated

ARRAY....continued

How to create a POLAR ARRAY

Using "Angle Between".

1. Draw a 3" Radius circle.

2. Add a .50 Radius 3 sided Polygon and place as shown.

3. Select the **ARRAY** command using one of the following:

   Ribbon = Home tab / Modify panel / Array ▼
   or
   Keyboard = Array <enter>

4. Select **Polar Array**.

5. Select Objects: *Select the Object to be Arrayed. (Polygon)*

6. Select Objects: *Select more objects or <enter> to stop*

7. Specify center point of array or [Base point / Axis of Rotation] **Select the Center Point of the Circle**

7. Snap to the center of the Circle to select the center of the Array

The **Array Creation** tab appears and the array defaults to 6 items.

8. Enter Items: **6**

9. Enter Between: **45**

10. Press <**enter**> to display the selections

11. Select **Close Array** if display is correct

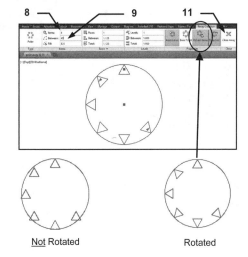

Note:

6 items were copied at each 45 degree ccw

Not Rotated   Rotated

1-9

ARRAY....continued

How to create a POLAR ARRAY

Using "Fill Angle".

1. Draw a 3" Radius circle.
2. Add a .50 Radius 3 sided Polygon and place as shown.
3. Select the **ARRAY** command using one of the following:
   Ribbon = Home tab / Modify panel / Array ▼
   or
   Keyboard = Array <enter>
4. Select **Polar Array**.
5. Select Objects: *Select the Object to be Arrayed. (Polygon)*
6. Select Objects: *Select more objects or <enter> to stop*
7. Specify center point of array or [Base point / Axis of Rotation] **Select the Center Point of the Circle**

       7. Snap to the center of the Circle to select the center of the Array

The **Array Creation** tab appears and the array defaults to 6 items.

8. Enter Items: **8**
9. Enter Fill: **180**
10. Press **<enter>** to display the selections
11. Select **Close Array** if display is correct

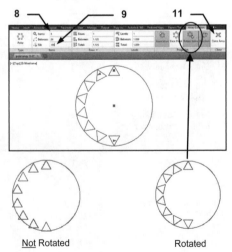

Note:

8 items were evenly distributed within 180 degrees

Not Rotated

Rotated

ARRAY....continued

How to create a PATH ARRAY

1. Draw a Line 6" long at 20 degrees..

2. Add a 0.500" x 0.500" Rectangle as shown.

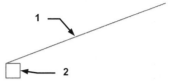

3. Select the **ARRAY** command using one of the following:

   Ribbon = Home tab / Modify panel / Array ▼
   or
   Keyboard = Array <enter>

4. Select **Path Array**.

5. Select Objects: *Select the Object to be Arrayed. (The small Rectangle)*

6. Select Objects: *Select more objects or <enter> to stop*

7. Specify Path Curve: *Select the Path. (The angled Line)*

> **Note:** The Path can be a line, polyline, spline, helix, arc, circle or ellipse.

The **Array Creation** tab appears and the array defaults to 9 items.

8. Make any alterations and press <**enter**> to display.

9. If correct select **Close Array**.

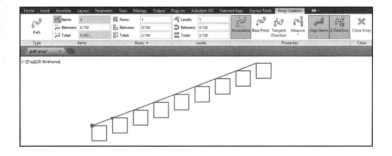

BREAK

The **BREAK** command allows you to break an object at a single point (**Break at Point**) or between two points (**Break**). I think of it as breaking a single line segment into two segments or taking a bite out of an object.

METHOD 1 - Break at a Single Point

How to break one Line into two separate objects with no visible space in between.

1. Select the **BREAK AT POINT** command by using:

   Ribbon = Home tab / Modify panel ▼ /

2. _break Select objects: *select the object to break (P1).*

3. Specify first break point: *select break location (P2) accurately.*

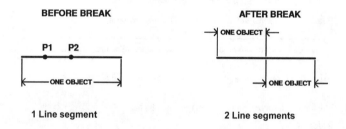

BEFORE BREAK

P1   P2

ONE OBJECT

1 Line segment

AFTER BREAK

ONE OBJECT

ONE OBJECT

2 Line segments

Note:

The single line is now 2 lines but no gap in between the 2 lines. For example, a 2 inch long line would become two 1 inch lines butted together.

Continued on the next page...

BREAK....continued

<u>METHOD 2 - Break between 2 points.</u> (Take a bite out of an object)

Use this method if the <u>location</u> of the BREAK <u>is not important</u>.

1. Select the **BREAK** command by using one of the following:

   Ribbon = Home tab / Modify panel ▼ /
   or
   Keyboard = BR <enter>

2. _break Select objects: *select the object to break (P1).*
3. Specify first break point: *select break location (P2) accurately.*

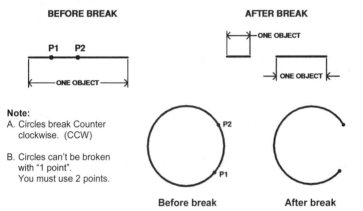

BEFORE BREAK   **AFTER BREAK**

P1   P2

ONE OBJECT

ONE OBJECT

ONE OBJECT

Note:

A. Circles break Counter
   clockwise. (CCW)

B. Circles can't be broken
   with "1 point".
   You must use 2 points.

P2

P1

Before break   After break

**The following method is the same as method 2 above; however, <u>use this method
if the location of the break is very specific.</u>**

1. Select the **BREAK** command.
2. _break Select objects: *select the object to break (P1) anywhere on the object.*
3. Specify second break point or [First point]: *type F <enter>.*
4. Specify first break point: *select the first break location (P2) accurately.*
5. Specify second break point: *select the second break location (P3) accurately.*

BEFORE BREAK   **AFTER BREAK**

P1 P2   P3   ONE OBJECT

ONE OBJECT   break

ONE OBJECT

CHAMFER

The **CHAMFER** command allows you to create a chamfered corner on two lines.
There are two methods: **Distance (below) and Angle (next page).**

DISTANCE METHOD

Distance Method requires input of a distance for each side of the corner.

1. Select the **CHAMFER** command using one of the following:

Ribbon = Home tab / Modify panel /
or
Keyboard = CHA <enter>

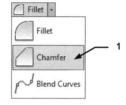

Command: _chamfer
(TRIM mode) Current chamfer Dist1 = 0.000, Dist2 = 0.000
Select first line or [Undo/Polyline/Distance/Angle/Trim/mEthod/Multiple]:

select "D" <enter>

Specify first chamfer distance <0.000>: *type the distance for first side <enter>.*
Specify second chamfer distance <0.000>: *type the distance for second side <enter>*

2. **Now chamfer the object.**
   Select first line or [Undo/Polyline/Distance/Angle/Trim/mEthod/Multiple]: *select the (First side) to be chamfered (distance 1).*
   Select second line or shift-select to apply corner or [Distance/Angle/Method]: *select the (Second side) to be chamfered (distance 2).*

Note: When you place the cursor on the second side, AutoCAD displays the Chamfer and allows you to change the Distances before it is actually drawn. If you choose to change the Distance, select the Distance option, enter new distance values then select the 2nd side.

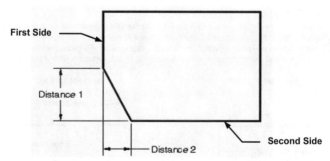

1-14

CHAMFER....continued

ANGLE METHOD

Angle method requires input for the length of the line and an angle.

1. Select the **CHAMFER** command

Command: _chamfer
(TRIM mode) Current chamfer Dist1 = 1.000, Dist2 = 1.000
Select first line or [Undo/Polyline/Distance/Angle/Trim/method/Multiple]: *type A <enter>*
Specify chamfer length on the first line <0.000>: *type the chamfer length <enter>*
Specify chamfer angle from the first line <0>: *type the angle <enter>*

2. **Now Chamfer the object**
   Select first line or [Undo/Polyline/Distance/Angle/Trim/mEthod/Multiple]: *select the (First Line) to be chamfered. (the length side)*
   Select second line or shift-select to apply corner: *select the (second line) to be chamfered. (the Angle side)*

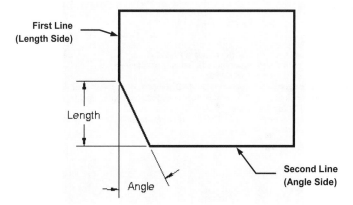

OPTIONS:

Polyline: This option allows you to Chamfer all intersections of a Polyline in one operation. Such as all 4 corners of a rectangle.

Trim: This option controls whether the original lines are trimmed or remain after the corners are chamfered. (Set to Trim or No trim.)

mEthod: Allows you to switch between **Distance** and **Angle** method. The distance or angle must have been set previously.

Multiple: Repeats the Chamfer command until you press **<enter>** or **Esc** key.

COPY

The **COPY** command creates a duplicate set of the objects selected.
The <u>COPY</u> command is similar to the <u>MOVE</u> command.

<u>The steps required are:</u>

1. Select the <u>objects to be copied</u>.
2. Select a <u>base point</u>.
3. Select a <u>New location for the New copy</u>.

<u>The difference between Copy and Move commands:</u>

The <u>Move</u> command <u>merely moves</u> the objects to a new location.
The <u>Copy</u> command <u>makes a copy</u> and you select the location for the new copy.

1. Select the **Copy** command using one of the following commands:

> **Ribbon = Home tab / Modify panel /**
> **or**
> **Keyboard = CO <enter>**

2. The following will appear on the command line:

Command: _copy
Select objects: *select the objects you want to copy*
Select objects: *stop selecting objects by selecting <enter>*
Current settings: Copy mode = Multiple
Specify base point or [Displacement/mOde] <Displacement>: *select a base point (P1)*
Specify second point of displacement or <use first point as displacement>: *select the new location (P2) for the first copy*
Specify second point or [Exit / Undo] <Exit>: *select the new location (P2) for the next copy or press <enter> to exit.*

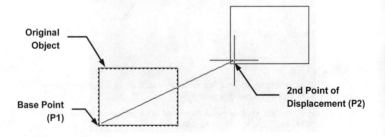

Original
Object

Base Point
(P1)

2nd Point of
Displacement (P2)

Continued on the next page...

COPY....continued

If you select the option **mOde**, you may select <u>Single</u> or <u>Multiple</u> copy mode. The default setting is Multiple. It is practical to leave the mode setting at Multiple. If you choose to make only one copy just press **<enter>** to exit the Copy command.

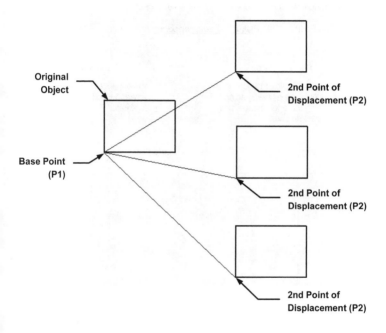

Original
Object

2nd Point of
Displacement (P2)

Base Point
(P1)

2nd Point of
Displacement (P2)

2nd Point of
Displacement (P2)

The copy command continues to make copies until you press **<enter>** to exit.

COPY "Array" option

The Copy command allows you to make an **Array** of copies.

After you have selected the Base point the following prompt appears:

Specify second point or [Array] <use first point as displacement>:

If you select the option **Array**,

1. Enter the number of items to Array: type in **4 <enter>**

2. Place 2nd Point or [Fit]: **Place 2nd point or select F <enter>**

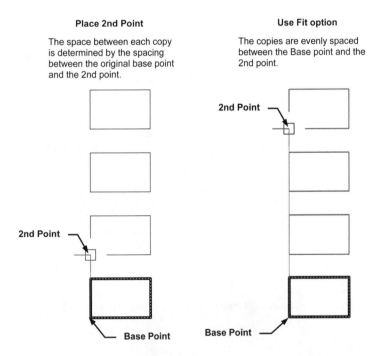

Place 2nd Point

The space between each copy is determined by the spacing between the original base point and the 2nd point.

2nd Point

Base Point

Use Fit option

The copies are evenly spaced between the Base point and the 2nd point.

2nd Point

Base Point

Note: The Array option within the Copy command is a quick method to create multiple copies. For more accurate array options use the Array command.

COPY using DRAG

The **Drag** option allows you to quickly **move** or **copy** an object(s).

EXAMPLE:
1. Draw a Circle.

2. Select the Circle.
   5 little boxes appear. These are Grips and allow you to edit the object.
   Grips will be discussed more in future lessons.

3. Click on the Circle and hold the right hand mouse button down as you drag the Circle to the right.

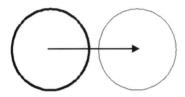

4. When the dragged Circle is in the desired location release the Right Mouse button and an options menu will appear.

5. Select **Move Here**, **Copy Here**, or **Cancel**.

Move Here: The original object selected will move to the new location.

Copy Here: The original object will remain in it's original location and a copy will appear in the new location.

DIVIDE

The **DIVIDE** command divides an object mathematically by the NUMBER of segments you specify. A POINT (object) is placed at each interval on the object.

Note: the object selected is **NOT** broken into segments. The **POINTS** are simply drawn **ON** the object.

EXAMPLE:

> This LINE has been DIVIDED into 4 EQUAL lengths.
> But remember, the line is not broken into segments.
> The Points are simply drawn **ON** the object.

1. First open the Point Style box and select the **POINT STYLE** to be placed on the object.

   Ribbon = Home tab / Utilities Panel / ▼ / Point Style
   or
   Keyboard = ddptype <enter>

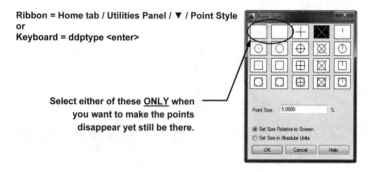

Select either of these <u>ONLY</u> when you want to make the points disappear yet still be there.

2. Next select the **DIVIDE** command using one of the following:

   Ribbon = Home tab / Draw Panel / ▼ /
   or
   Keyboard = DIV <enter>

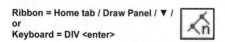

3. Select Object to divide: *select the object to divide*

4. Enter the number of segments or [Block]: *type the number of segments <enter>*

ERASE

There are 3 methods to erase (delete) objects from the drawing.
They all work equally well. You decide which one you prefer to use.

Method 1
Select the **Erase** command first and then select the objects.

Example:
1. Start the **Erase** command using one of the following:

Ribbon = Home tab / Modify panel /
or
Keyboard = E <enter>

2. Select objects: *Pick one or more objects*
   Select Objects: *Press <enter> and the objects selected will disappear.*

Method 2
Select the Objects first and then the **Delete** Key.

Example:
1. Select the object to be erased.
2. Press the **Delete** Key.

Method 3
Select the Objects first and then select the **Erase** command from the Shortcut Menu.

Example:
1. Select the object to be erased.
2. Press the right Mouse button.
3. Select **Erase** from the Shortcut Menu
   using the left mouse button.

3 ——

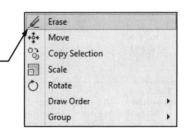

Note: Very Important
If you want the erased objects to return, select the **Undo tool** from the **Quick Access Toolbar**. This will Undo the last command.
More about Undo and Redo later.

EXPLODE

The **EXPLODE** command changes (explodes) an object into its primitive objects.
For example: A rectangle is originally one object. If you explode it, it changes into 4
lines. Visually you will not be able to see the change unless you select one of the lines.

1. Select the **Explode** command by using one of the following:

   Ribbon = Home tab / Modify panel /
   or
   Keyboard = X <enter>

2. The following will appear on the command line:

   Command: _explode
   Select objects: *select the object(s) you want to explode.*
   Select objects: *select <enter>.*

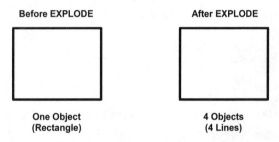

Before EXPLODE	**After EXPLODE**
One Object (Rectangle)	4 Objects (4 Lines)

**(Notice there is no visible difference. But now you have 4 lines instead of 1
Rectangle).**

Try this:
Draw a rectangle and then click on it. The entire object highlights. Now explode the
rectangle, then click on it again. Only the line you clicked on should be highlighted.
Each line that forms the rectangular shape is now an individual object.

EXTEND

The **EXTEND** command is used to extend an object to a **boundary**. The object to be extended must actually or theoretically intersect the boundary.

1. Select the **EXTEND** command using one of the following:

Ribbon = Home tab / Modify panel /
or
Keyboard = EX <enter>

2. The following will appear on the command line:

Command: _extend
Current settings: Projection = UCS   Edge = Extend
Select boundary edges ...
Select objects or <select all>: *select boundary (P1) by clicking on the object.*
Select objects: *stop selecting boundaries by selecting <enter>.*
Select object to extend or shift-select to Trim or
[Fence/Crossing/Project/Edge/Undo]: *select the object that you want to extend (P2 and P3). (Select the end of the object that you want to extend.)*
Select object to extend or [Fence/Crossing/Project/Edge/Undo]: *stop selecting objects by pressing <enter>.*

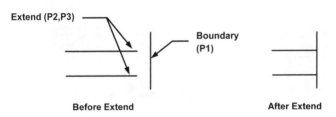

Before Extend   **After Extend**

Note: When selecting the object to be extended (**P2** and **P3** above) click on the end pointing towards the boundary.

Fence Use a "Fence" line to select objects to extend
Crossing You may select objects using a Crossing Window
Project Same as Edge except used only in "3D".

Edge (Extend or No Extend)
   In the **"Extend"** mode, (default mode) the boundary and the Objects to be extended need only intersect if the objects were infinite in length.
   In the **"No Extend"** mode the boundary and the objects to be extended must visibly intersect.

Undo You may **"undo"** the last extended object while in the Extend command

FILLET

The **FILLET** command will create a radius between two objects. The objects do not have to be touching. If two parallel lines are selected, it will construct a full radius.

RADIUS A CORNER

1. Select the **FILLET** command using one of the following:

    Ribbon = Home tab / Modify panel /
    or
    Keyboard = F <enter>

2. The following will appear on the command line:

3. **Set the radius of the fillet**
    Command: _fillet
    Current settings: Mode = TRIM, Radius = 0.000
    Select first object or [Undo/Polyline/Radius/Trim/Multiple]: *type "R" <enter>*
    Specify fillet radius <0.000>: *type the radius <enter>*

4. **Now fillet the objects**
    Select first object or [Undo/Polyline/Radius/Trim/Multiple]: *select the 1st object to*
    be filleted
    Select second object or shift-select to apply corner or [Radius]: *select the 2nd*
    object to be filleted

Note: When you place the Cursor on the second object, AutoCAD displays the Fillet and allows you to change the Radius before it is actually drawn. If you choose to change the Radius, select the Radius option, enter a new radius value then select the 2nd object.

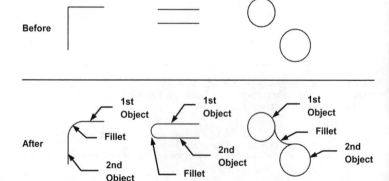

FILLET....continued

The **FILLET** command may also be used to create a square corner.

SQUARE CORNER

1. Select the **FILLET** command

2. The following will appear on the command line:

Select first object or [Undo/Polyline/Radius/Trim/Multiple]: *select the 1st object (P1)*

Select second object or shift-select to apply corner: *Hold the shift key down while selecting the 2nd object (P2)*

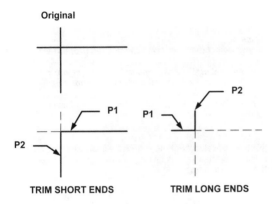

TRIM SHORT ENDS   **TRIM LONG ENDS**

Note: The corner trim direction depends on which end of the object you select. Select the ends that you wish to keep.

OPTIONS:

Polyline: This option allows you to fillet all intersections of a Polyline in one operation, such as all 4 corners of a rectangle.

Trim: This option controls whether the original lines are trimmed to the end of the Arc or remain the original length. (Set to **Trim** or **No trim**)

Multiple: Repeats the fillet command until you press <enter> or **Esc** key.

MATCH PROPERTIES

Match Properties is used to **"paint"** the properties of one object to another. This is a simple and useful command. You first select the object that has the desired properties (the source object) and then select the object you want to "paint" the properties to (destination object).

Only one **"source object"** can be selected but its properties can be painted to any number of "destination objects".

1. Select the Match Properties command using one of the following:

   Ribbon = Home tab / Clipboard panel /
   or
   Keyboard = MA <enter>

   Command: matchprop
2. Select source object: *select the object with the desired properties to match*

3. Select destination object(s) or [Settings]: *select the object(s) you want to receive the matching properties.*

4. Select destination object(s) or [Settings]: *select more objects or <enter> to stop.*

Note: If you do not want to match all of the properties, after you have selected the source object, right click and select **"Settings"** from the short cut menu, before selecting the destination object. Uncheck all the properties you do not want to match and select the **OK** button. Then select the destination object(s).

MEASURE

The **MEASURE** command is very similar to the **DIVIDE** command because point objects are drawn at intervals on an object. However, the **MEASURE** command allows you to specify the **LENGTH** of the segments rather than the number of segments.

Note: the object selected is **NOT** broken into segments. The **POINTS** are simply drawn **ON** the object.

EXAMPLE:

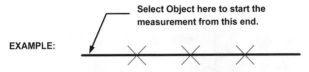

The **MEASURE**ment was started at the left endpoint, and ended just short of the right end of the line. The remainder is less than the measurement length specified.
You designate which end you want the measurement to start by selecting the end when prompted to select the object.

1. First open the Point Style box and select the **POINT STYLE** to be placed on the object.

   Ribbon = Home tab / Utilities Panel ▼/ Point Style
   or
   Keyboard = ddptype <enter>

   (Refer to the **Point** command if you need a refresher on Points)

2. Next select the **MEASURE** command using one of the following:

   Ribbon = Home tab / Draw Panel ▼/
   or
   Keyboard = ME <enter>

3. Select Object to MEASURE: *select the object to MEASURE*
   (Note: this selection point is also where the MEASUREment will start.)

4. Specify length of segment or [Block]: *type the length of one segment <enter>*

MEASURE TOOLS and ID POINT

The following tools are very useful to confirm the location or size of objects.

The Measure tools enables you to measure the Distance, Radius, Angle, Area, or Volume of a selected object. The default option is Distance.

1. You may access these tools as follows:

Ribbon = Home tab / Utilities Panel / Measure ▼ ▢

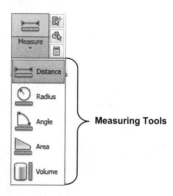

Measuring Tools

2. Select one of the tools and follow the instructions on the command line.

ID Point

The ID Point command will list the X and Y coordinates of the point that you select. The coordinates listed will be from the Origin.

Ribbon = Home tab / Utilities Panel / ▼
or
Keyboard = ID <enter>

1. Select the ID Point command by typing: **ID <enter>**
2. Select a location point, such as the endpoint of a line.
   The X, Y and Z location coordinates for the endpoint will be displayed.

Example:
1. Command: **id<enter>**
2. Snap to the endpoint
3. Coordinates, from the Origin, are displayed.

2. Snap to endpoint

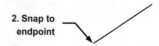

Command: ' id Specify point: X = 5.474   Y = 0.791   Z = 0.000

MIRROR

The **MIRROR** command allows you to make a mirrored image of any objects you select. You can use this command for creating right / left hand parts or draw half of a symmetrical object and mirror it to save drawing time.

1. Select the **MIRROR** command using one of the following:

   Ribbon = Home tab / Modify Panel /
   or
   Keyboard = MI <enter>

2. The following will appear on the command line:

   Command: _mirror
   Select objects: *select the objects to be mirrored*
   Select objects: *stop selecting objects by selecting <enter>*
   Specify first point of mirror line: *select the 1st end of the mirror line*
   Specify second point of mirror line: *select the 2nd end of the mirror line*
   Erase source objects? [Yes/No] <N>: **select Y or N**

1st end of Mirror Line

Original Object

Mirrored Image

2nd end of Mirror Line (Vertical)

Mirror Line "Vertical"

Continued on the next page...

MIRROR....continued

Note:
The placement of the **"Mirror Line"** is important. You may make a mirrored copy
horizontally, vertically or on an angle. See examples below and on the previous page.

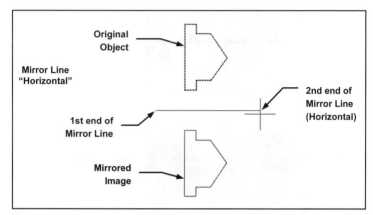

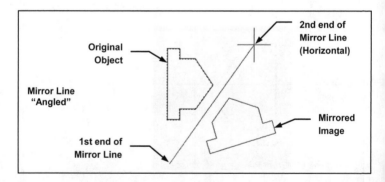

How to control text when using the Mirror command:
(Do the following **before** you use the mirror command)

1. At the command line type: *mirrtext <enter>*
2. If you want the text to mirror (reverse reading) type: *1 <enter>*
   If you do not want the text to mirror, type: *0 <enter>*

   ꓩ = ⅁ИITT3Ƨ TX3TЯЯIM   MIRRTEXT SETTING = 0

MOVE

The **MOVE** command is used to move object(s) from their current location (base point) to a new location (second displacement point).

1. Select the Move command using one of the following:

Ribbon = Home tab / Modify Panel /
or
Keyboard = M <enter>

2. The following will appear on the command line:

Command: _move
Select objects: *select the object(s) you want to move (P1).*
Select objects: *select more objects or stop selecting object(s) by selecting <enter>*
Specify base point or displacement: *select a location (P2) (usually on the object).*
Specify second point of displacement or <use first point as displacement>: *move the object to its <u>new location</u> (P3) and press the left mouse button.*

Warning: If you press <enter> instead of actually picking a new location (P3) with the mouse, AutoCAD will send it into Outer Space. If this happens just select the undo tool and try again.

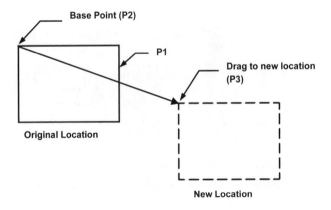

MOVE using DRAG

The **Drag** option allows you to quickly **move** or **copy** an object(s).

EXAMPLE:
1. Draw a Circle.

2. Select the Circle.
   5 little boxes appear. These are Grips and allow you to edit the object.
   Grips will be discussed more in future lessons.

3. Click on the Circle and hold the right hand mouse button down as you drag the
   Circle to the right.

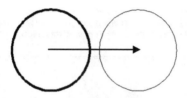

4. When the dragged Circle is in the desired location release the Right Mouse button
   and an options menu will appear.

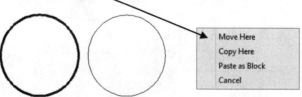

5. Select **Move Here**, **Copy Here**, or **Cancel**.

Move Here: The original object selected will move to the new location.

Copy Here: The original object will remain in it's original location and a copy will
            appear in the new location.

NUDGE

The **Nudge** option allows you to nudge objects in orthogonal increments.

Note:
Snap mode affects the distance and direction in which the objects are nudged.

Nudge objects with **Snap** mode turned **OFF**:
Objects move two pixels at a time.

Nudge objects with **Snap** mode turned **ON**:
Objects are moved in increments specified by the current **Snap** spacing.

EXAMPLE:

1. Draw a Circle.

2. Select the Circle.
   5 little boxes appear. These are Grips and allow you to edit the object.
   Grips will be discussed more in future lessons.

3. Hold down the **Ctrl** key and press one of the **Arrow** keys

Remember:
The distance the object moves depends on whether you have the **Snap** mode **ON** or
OFF. (Refer to note above)

OFFSET

The **OFFSET** command duplicates an object parallel to the original object at a specified distance. You can offset Lines, Arcs, Circles, Ellipses, 2D Polylines and Splines. You may duplicate the original object or assign the offset copy to another layer.

Examples of Offset objects.

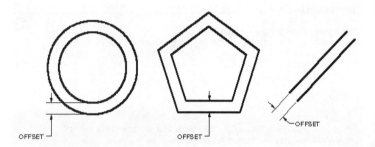

HOW TO USE THE OFFSET COMMAND:

METHOD 1
(Duplicate the Original Object)

1. Select the **OFFSET** command using one of the following:

    Ribbon = Home tab / Modify Panel /
    or
    Keyboard = offset <enter>

2. Command: _offset
    Current settings: Erase source=No Layer=Source OFFSETGAPTYPE=0
    Specify offset distance or [Through/Erase/Layer] <Through>: *type the offset distance or select Erase or Layer. (see options on the next page)*

3. Select object to offset or <Exit/Undo>: *select the object to offset*.

4. Specify point on side to offset or [Exit/Multiple/Undo]<Exit>: *Select which side of the original you want the duplicate to appear by placing your cursor and clicking. (See options on the next page)*

5. Select object to offset or [Exit/Undo]<Exit>: *Press <enter> to stop*.

OFFSET....continued

METHOD 2
(Duplicate original object but assign the Offset copy to a different layer)

To automatically place the <u>offset copy</u> on a different layer than the original you must first change the "current" layer to the layer you want the offset copy to be placed on.

1. Select the layer that you want the offset copy placed on from the list of layers.

2. Select the **OFFSET** command (refer to previous page)

3. Command: _offset
   Current settings: Erase source=No Layer=Source OFFSETGAPTYPE=0
   Specify offset distance or [Through/Erase/Layer] <Through>: *type L <enter>*

4. Enter layer option for offset objects [Current/Source] <Source>: *select C <enter>*

5. Specify offset distance or [Through/Erase/Layer] <Through>: *type the offset distance <enter>*
6. Select object to offset or [Exit/Undo] <Exit>: *select the object to offset.*

7. Specify point on side to offset or [Exit/Multiple/Undo]<Exit>: *Select which side of the object you want the duplicate to appear by placing your cursor and clicking. (See options below)*

8. Select object to offset or [Exit/Undo]<Exit>: *Press <enter> to stop.*

OPTIONS:

Through: Creates an object passing through a specified point.

Erase: Erases the source object after it is offset.

Layer: Determines whether offset objects are created on the <u>current</u> layer or on the layer of the <u>source</u> object. Select <u>Layer</u> and then select <u>current</u> or <u>source</u>. (Source is the default)

Multiple: Turns on the multiple offset mode, which allows you to continue creating duplicates of the original without re-selecting the original.

Exit: Exits the Offset command.

Undo: Removes the previous offset copy.

ROTATE

The **ROTATE** command is used to rotate objects around a Base Point. (pivot point)
After selecting the objects and the base point, you will enter the rotation angle from its
<u>current</u> rotation angle or select a reference angle followed by the new angle.

A **Positive** rotation angle revolves the objects **Counter-Clockwise**.
A **Negative** rotation angle revolves the objects **Clockwise**.

Select the **ROTATE** command using one of the following:

> **Ribbon = Home tab / Modify Panel /**
> **or**
> **Keyboard = RO <enter>**

ROTATION ANGLE OPTION
Command: _rotate
1. Current positive angle in UCS: ANGDIR=counterclockwise ANGBASE=0
   Select objects: *select the object to rotate.*
2. Select objects: *select more object(s) or <enter> to stop.*
3. Specify base point: *select the base point (pivot point).*
4. Specify rotation angle or [Copy/Reference]<0>: *type the angle of rotation.*

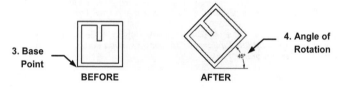

REFERENCE OPTION
Command: _rotate
1. Current positive angle in UCS: ANGDIR=counterclockwise ANGBASE=0
   Select objects: *select the object to rotate.*
2. Select objects: *select more object(s) or <enter> to stop.*
3. Specify base point: *select the base point (pivot point).*
4. Specify rotation angle or [Reference]: *select Reference.*
5. Specify the reference angle <0>: *Snap to the reference object (1) and (2).*
6. Specify the new angle or [Points]: *P <enter>.*
7. Specify first point: *select 1st endpoint of new angle.*
8. Specify second point: *select 2nd endpoint of new angle.*

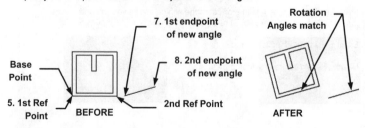

SCALE

The **SCALE** command is used to make objects larger or smaller proportionately. You may scale using a scale factor or a reference length. You must also specify a base point. The base point is a stationary point from which the objects scale.

1. Select the **SCALE** command using one of the following:

> **Ribbon = Home tab / Modify Panel /**
> or
> **Keyboard = scale <enter>**

SCALE FACTOR

Command: _scale
2. Select objects: *select the object(s) to be scaled*
3. Select objects: *select more object(s) or <enter> to stop*
4. Specify base point: *select the stationary point on the object*
5. Specify scale factor or [Copy/Reference]: *type the <u>scale factor</u> <enter>*

If the scale factor is greater than 1, the objects will increase in size.
If the scale factor is less than 1, the objects will decrease in size.

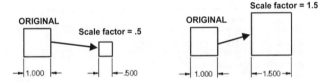

REFERENCE option

Command: _scale
2. Select objects: *select the object(s) to be scaled*
3. Select objects: *select more object(s) or <enter> to stop*
4. Specify base point: *select the stationary point on the object*
5. Specify scale factor or [Copy/Reference]: *select Reference*
6. Specify reference length <1>: *specify a <u>reference</u> length*
7. Specify new length: *specify the new length*

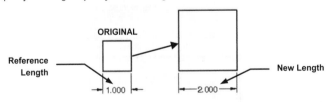

<u>COPY</u> option - creates a duplicate of the selected object. The duplicate is directly on top of the original. The duplicate will be scaled. The Original remains the same.

STRETCH

The **STRETCH** command allows you to stretch or compress object(s). Unlike the Scale command, you can alter an objects proportion with the Stretch command. In other words, you may increase the length without changing the width and vice versa.

Stretch is a very valuable tool. Take some time to really understand this command. It will save you hours when making corrections to drawings.

When selecting the object(s) you must use a **CROSSING** window.
Objects that are crossed, will *stretch*.
Objects that are totally enclosed, will *move*.

1. Select the **STRETCH** command using one of the following:

> **Ribbon = Home tab / Modify Panel /**
> **or**
> **Keyboard = S <enter>**

Command: _stretch
2. Select objects to stretch by crossing-window or crossing-polygon...
   Select objects: *select the first corner of the crossing window*
3. Specify opposite corner: *specify the opposite corner of the crossing window*
4. Select objects: *<enter>*
5. Specify base point or [Displacement] <Displacement>:
   select a base point (where it stretches from)
6. Specify second point or <use first point as displacement>:
   type coordinates or place location with cursor

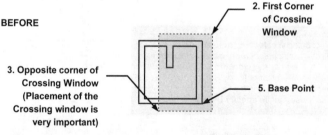

BEFORE

2. First Corner of Crossing Window

3. Opposite corner of Crossing Window (Placement of the Crossing window is very important)

5. Base Point

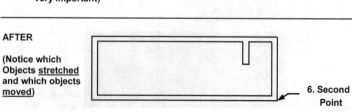

AFTER

(Notice which Objects <u>stretched</u> and which objects <u>moved</u>)

6. Second Point

TRIM

The **TRIM** command is used to trim an object to a **cutting edge**. You first select the "Cutting Edge" and then select the part of the object you want to trim. The object to be trimmed must actually intersect the cutting edge or <u>could</u> intersect if the objects were infinite in length.

1. Select the Trim command using one of the following:

Ribbon = Home tab / Modify Panel /
or
Keyboard = TR <enter>

2. The following will appear on the command line:

Command: _trim
Current settings: Projection = UCS   Edge = Extend
Select cutting edges ...
Select objects or <select all>: *select cutting edge(s) by clicking on the object (P1)*
Select objects: *stop selecting cutting edges by pressing the <enter> key*
Select object to trim or shift-select to extend or
[Fence/Crossing/Project/Edge/eRase/Undo]: *select the object that you want to trim.*
(P2) (Select the part of the object that you want to disappear, not the part you
want to remain)
Select object to trim or [Fence/Crossing/Project/Edge/eRase/Undo]: *press <enter> to*
stop

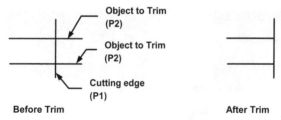

Object to Trim
(P2)

Object to Trim
(P2)

Cutting edge
(P1)

Before Trim   **After Trim**

Note: You may toggle between Trim and Extend. Hold down the **SHIFT** key and the Extend command will activate. Release the **SHIFT** key and you return to Trim.

Fence Use a "Fence" line to select objects to extend.

Edge See page 1-23

Project Same as Edge except used only in "3D"

Crossing You may select objects using a Crossing Window.

eRase You may erase an object instead of trimming while in the Trim command.

Undo You may "undo" the last trimmed object while in the Trim command

UNDO and REDO

The **UNDO** and **REDO** tools allow you to undo or redo <u>previous commands</u>.
For example, if you erase an object by mistake, you can UNDO the previous "erase"
command and the object will reappear. So don't panic if you do something wrong. Just
use the UNDO command to remove the previous commands.

The **Undo** and **Redo** tools are located in the **Quick Access Toolbar**.

Note:
You may **UNDO** commands used during a work session until you close the drawing.

How to use the <u>Undo</u> tool.

1. Draw a line, circle and a rectangle.

Your drawing should look approximately like this.

2. Next Erase the Circle and the Rectangle.

(The Circle and the Rectangle disappear.)

3. Select the UNDO arrow.

You have now deleted the ERASE command operation.
As a result the erased objects reappear.

How to use the <u>Redo</u> command:
Select the REDO arrow and the Circle and Rectangle will disappear again.

WIPEOUT

The Wipeout command creates a blank area that covers existing objects. The area has a background that matches the background of the drawing area. This area is bounded by the wipeout frame, which you can turn on or off.

1. Select the Wipeout command using one of the following:

Ribbon = Home tab / Draw Panel ▼ /
or
Keyboard = Wipeout <enter>

.   Command: _wipeout Specify first point or [Frames/Polyline] <Polyline>: ***specify the***
                                                                first point of the shape (P1)
3. Specify next point: ***specify the next point (P2)***
4. Specify next point or [Undo]: ***specify the next point (P3)***
5. Specify next point or [Undo]: ***specify the next point or <enter> to stop***

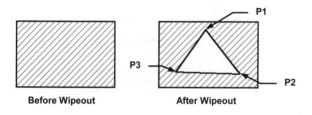

Before Wipeout   **After Wipeout**

TURNING FRAMES ON OR OFF

1. Select the Wipeout command.
2. Select the **"Frames"** option.
3. Enter **ON** or **OFF**.

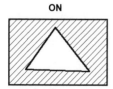
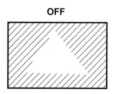

ON   OFF

Note: If you want to move the objects and the wipeout area, you must select both and move them at the same time. Do not move them separately.

ZOOM

The **ZOOM** command is used to move closer to or farther away from an object.
This is called Zooming In and Out.

1. Select the Zoom command by using the following:

Ribbon = View tab / Navigate 2D panel

2. Select the ▼ down arrow to display all of the selections.

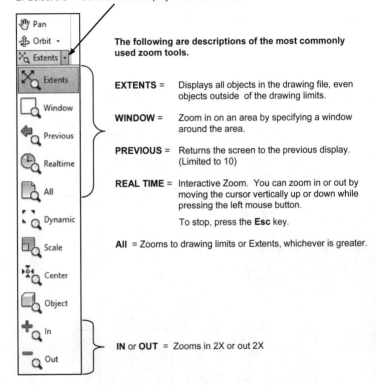

The following are descriptions of the most commonly used zoom tools.

EXTENTS = Displays all objects in the drawing file, even objects outside of the drawing limits.

WINDOW = Zoom in on an area by specifying a window around the area.

PREVIOUS = Returns the screen to the previous display. (Limited to 10)

REAL TIME = Interactive Zoom. You can zoom in or out by moving the cursor vertically up or down while pressing the left mouse button.

To stop, press the **Esc** key.

All = Zooms to drawing limits or Extents, whichever is greater.

IN or **OUT** = Zooms in 2X or out 2X

You may also select the Zoom commands using one of the following:

Right Click and select Zoom from the Short cut menu.

Keyboard = Z <enter> Select from the options listed.

ZOOM....continued

How to use ZOOM / WINDOW

1. Select Zoom / Window (Refer to previous page)

2. Create a window around the objects you want to enlarge.
   (Creating a "window" is a similar process to drawing a rectangle. It requires a first corner and then a diagonal corner)

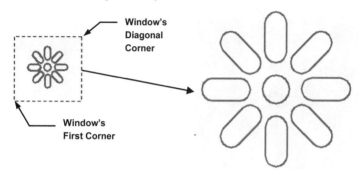

Window's Diagonal Corner

Window's First Corner

<u>Magnified to this view</u>

Note: the objects have been magnified. But the actual size has not changed.

How to return to Original View

1. Type: **Z** <enter> **A** <enter> (This is a shortcut for Zoom / All)

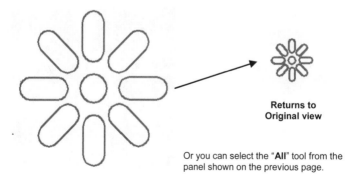

Returns to Original view

Or you can select the "**All**" tool from the panel shown on the previous page.

Notes:

Section 2
Concepts

MODEL and LAYOUT OPTIONS

Very important:

*Before I discuss Model and Layout I need you to confirm **Model and Layout tabs** are displayed.*

This will just take a minute.

1. Type: **options <enter>**

2. Select the **Display** tab.

3. Check and un-check boxes as shown

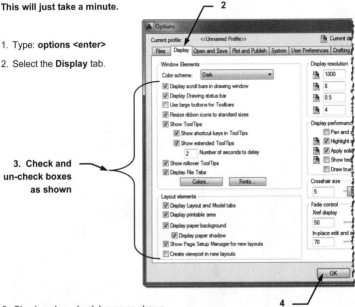

3. Check and un-check boxes as shown.

4. Select the **OK** button

5. The lower left corner of the drawing area should display the 3 tabs, Model, Layout1 and Layout2 and a few tools should be displayed in the lower right corner above the command line.

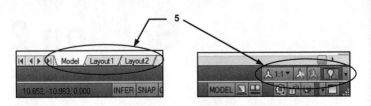

MODEL and LAYOUT Tabs

Read this information carefully. It is very important that you understand this concept. More information on the following pages.

AutoCAD provides two drawing spaces, **MODEL** and **LAYOUT**. You move into one or the other by selecting either the MODEL or LAYOUT tabs, located at the bottom left of the drawing area. (If you do not have these displayed follow the instructions on the previous page.)

Refer to the previous page if you do not have these

Model Tab (Also called *Model Space*)

When you select the Model tab you enter <u>MODEL SPACE</u>.
Model Space is where you **create** and **modify** your drawings.

Layout Tabs (Also called *Paper Space*)

When you select a Layout tab you enter <u>PAPER SPACE</u>.
The primary function of Paper Space is to **prepare the drawing for plotting**.

When you select the Layout tab for the first time, the "<u>Page Setup Manager</u>" dialog box will appear. The Page Setup Manager allows you to specify the printing device and paper size to use.

(More information on this in "How to …..)

When you select a Layout tab, Model Space will seem to have disappeared, and a <u>blank sheet of paper</u> is displayed on the screen. This sheet of paper is basically <u>in front of the Model Space</u>. (Refer to the illustration on the next page)

To see the drawing in Model Space, while still in Paper Space, you must <u>cut a hole</u> in this sheet. This hole is called a **"Viewport"**. *(Refer to "How to….)*

MODEL and LAYOUT Tabs....continued

Try to think of this as a picture frame (paper space) in front of a photograph (model space).

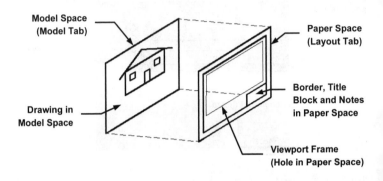

Model Space
(Model Tab)

Paper Space
(Layout Tab)

Drawing in
Model Space

Border, Title
Block and Notes
in Paper Space

Viewport Frame
(Hole in Paper Space)

This is what you see when you select the `Model` tab.

You see only Model Space.

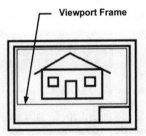

Viewport Frame

This is what you see when you select the `Layout1` tab with a Viewport.

You see through the Viewport to Model Space.

WHY LAYOUTS ARE USEFUL

I know you are probably wondering why you should bother with Layouts.
A Layout (Paper Space) is a great method to manipulate your drawing for plotting.

Notice the drawing below with <u>multiple viewports</u>.

Each viewport is a <u>hole</u> in the paper.
You can see through each viewport (hole) to model space.

Using Zoom and Pan you can manipulate the display of model space in each viewport.
To manipulate the display you must be <u>inside</u> the viewport.

Note: the dashed line indicates the
maximum printing area for the printer
and paper selected. Any object outside
of this area will not print.

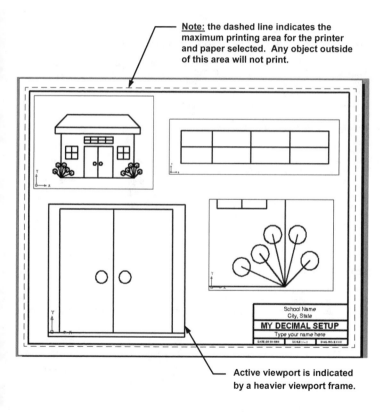

**Active viewport is indicated
by a heavier viewport frame.**

CREATING SCALED DRAWINGS

A very important rule in CAD you must understand is:

"*All objects are drawn full size*"

In other words, if you want to draw a line 20 feet long, you actually draw it 20 feet long. If the line is 1/8" long, you actually draw it 1/8" long.

Drawing and Plotting objects that are very large or very small.

What if you wanted to draw a house? Could you print it to scale on an 8-1/2 X 11 piece of paper? How about a small paper clip. Could you make it big enough to dimension? Let's start with the house.

How to print an entire house on an 8-1/2 X 11 sheet of paper.

Remember the photo and picture frame example I suggested on page 2-4. This time try to picture yourself standing at the front door of your house with an empty picture frame in your hands. Look at your house through the picture frame. Of course the house is way too big to fit in the frame. Or is it because you are standing too close to the house?

Now walk across the street and look through the picture frame in your hands again. Does the house appear smaller? Can you see all of it in the frame? If you could walk far enough away from the house it would eventually appear small enough to fit in the picture frame in your hands. But....the house did not actually change size, did it? It only appears smaller because you and the picture frame are farther away from it.

Adjusting the Viewport scale.

When using AutoCAD, walking across the street with the frame in your hands is called **Adjusting the Viewport scale**. You are increasing the distance between model space (your drawing) and Paper space (Layout) and that makes the drawing appear smaller.

For example: A viewport scale of 1/4" = 1' would make model space appear 48 times smaller. But, when you dimension the house, the dimension values will be the actual measurement of the house. In other words, a 30 ft. line will have a dimension of 30'-0".

When plotting something smaller, like a paperclip, you have to move the picture frame closer to model space to make it appear larger. **For example:** 8 = 1

ADJUSTING THE VIEWPORT SCALE

The following will take you through the process of adjusting the scale within a viewport.

1. **Open** a drawing.

2. Select a **Layout** tab. (paper space)

3. Cut a new Viewport or unlock an existing Viewport. (See "How to....")

4. **Zoom / All** to display all of the drawing limits.

5. Adjust the **Scale**.
   A. You must be in Paper Space. (See below)
   B. Select the Viewport Frame.
   C. Unlock Viewport, if locked.
   D. Select the Viewport Scale down arrow.
   E. Select the scale from the list of scales.

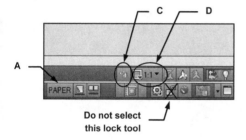

Do not select
this lock tool

6. Lock the Viewport.

Note: If you would like to add a scale that is on the list:

1. Type **Scalelistedit**
2. Select **Add** button.

3. Enter Scale name to display in scale list.
4. Enter Paper and drawing units.
5. Select the **OK** button.

2-7

Notes:

Section 3
Dimensioning

DIMENSIONING

Dimensions can be Associative, Non-Associative or Exploded. You need to understand what these are so you may decide which setting you want to use. Most of the time you will use Associative but you may have reasons to also use Non-Associative and Exploded.

Associative

Associative Dimensioning means that the dimension is actually associated to the objects that they dimension. If you move the object, the dimension will move with it. If you change the size of the object, the dimension text value will change also. (Note: This is not parametric. In other words, you cannot change the dimension text value and expect the object to change. That would be parametric dimensioning)

Non-Associative

Non-Associative means the dimension is not associated to the objects and will not change if the size of the object changes.

Exploded

Exploded means the dimension will be exploded into lines, text and arrowheads and non-associative.

How to set dimensioning to Associative, Non-Associative or Exploded.

1. On the command line type: *dimassoc <enter>*

2. Enter the number *2, 1 or 0 <enter>*

   2 = Associative

   1 = Non-Associative

   0 = Exploded

Note: The default factory setting is '2' **Associative**.

DIMENSIONING....continued

How to Re-associate a dimension.

If a dimension is <u>Non-associative</u>, and you would like to make it <u>Associative</u>, you may use the **dimreassociate** command to change.

1. Select **Reassociate** using one of the following:

   Ribbon = Annotate Tab / Dimension ▼ Panel /
   or
   Keyboard = dimreassociate <enter>

2. Select objects: *select the dimension to be reassociated*.

3. Select objects: *select more dimensions or <enter> to stop*.

4. Specify first extension line origin or [Select object] <next>: *(an "X" will appear to identify which is the first extension); use object snap to select the exact location, on the object, for the extension line point*.

5. Specify second extension line origin <next>: *(the "X" will appear on the second extension) use object snap to select the exact location, on the object, for the extension line point*.

6. *Continue until all extension line points are selected*.

<u>Note: You must use object snap to specify the exact location for the extension lines</u>.

Regenerating Associative dimensions

Sometimes after panning and zooming, the associative dimensions seem to be floating or not following the object. The **DIMREGEN** command will move the associative dimensions back into their correct location.

Type: *dimregen <enter>*

DIMENSION STYLES

Using the "Dimension Style Manager" you can change the appearance of the dimension features, such as length of arrowheads, size of the dimension text, etc. There are over 70 different settings. You can also Create New, Modify and Override Dimension Styles. All of these are simple, by using the Dimension Style Manager described below.

1. Select the "<u>Dimension Style Manager</u>" using one of the following:

Ribbon = Annotate Tab / Dimension Panel / ⬂
or
Keyboard = dimstyle <enter>

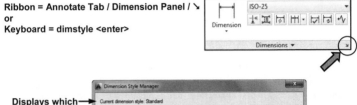

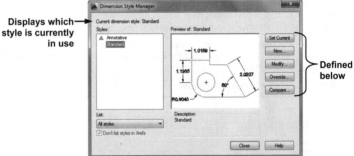

Displays which style is currently in use

Defined below

Set Current Select a style from the list of styles and select the **set current** button.

New Select this button to create a new style. When you select this button, the **Create New Dimension Style** dialog box is displayed.

Modify Selecting this button opens the **Modify dimension Style** dialog box which allows you to make changes to the style selected from the "list of styles".

Override An override is a temporary change to the current style. Selecting this button opens the Override Current Style dialog box.

Compare Compares two styles.

CREATING A NEW DIMENSION STYLE

A dimension style is a group of settings that has been saved with a name you assign. When creating a new style you must start with an existing style, such as Standard. Next, assign it a new name, make the desired changes and when you select the **OK** button the new style will have been successfully created.

How to create a NEW dimension style.

1. Open a drawing.

2. Select the **Dimension Style Manager** command (Refer to previous page)

3. Select the **NEW** button.

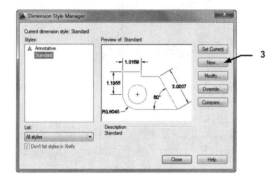

4. Enter **CLASS STYLE** in the **"New Style Name"** box.

5. Select **STANDARD** in the **"Start With:"** box.

6. **"Use For:"** box will be discussed later. For now, leave it set to **"All dimensions"**.

7. Select the **CONTINUE** button.

CREATING A NEW DIMENSION STYLE....continued

8. Select the **Primary Units** tab and change your settings to match the settings shown below.

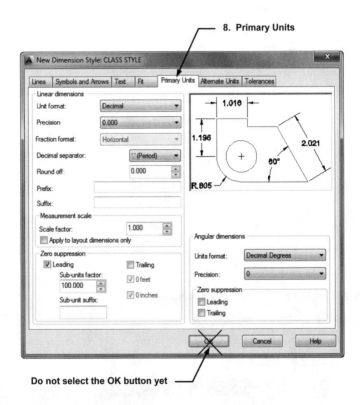

8. Primary Units

Do not select the OK button yet

CREATING A NEW DIMENSION STYLE....continued

9. Select the **Lines** tab and change your settings to match the settings shown below.

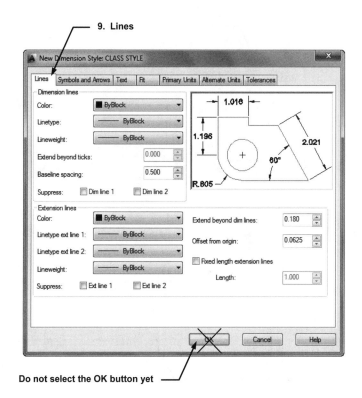

9. Lines

Do not select the OK button yet ———

CREATING A NEW DIMENSION STYLE....continued

10. Select the **Symbols and Arrows** tab and change your settings to match the settings shown below.

10. Symbols and Arrows

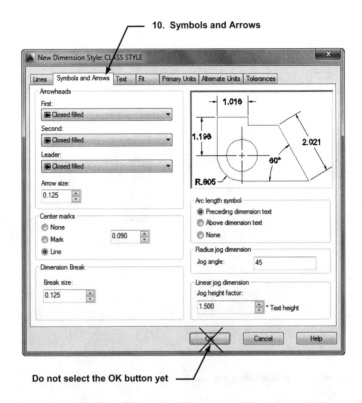

Do not select the OK button yet

CREATING A NEW DIMENSION STYLE....continued

11. Select the **Text** tab and change your settings to match the settings shown below.

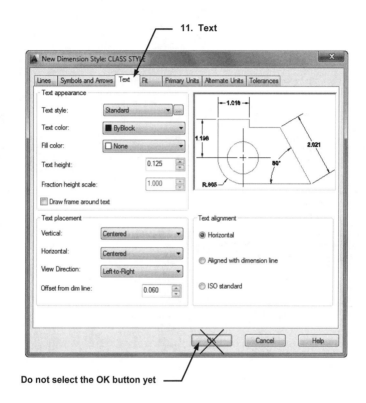

11. Text

Do not select the OK button yet

CREATING A NEW DIMENSION STYLE....continued

12. Select the **Fit** tab and change your settings to match the settings shown below.

13. Now select the **OK** button.

12. **Fit**

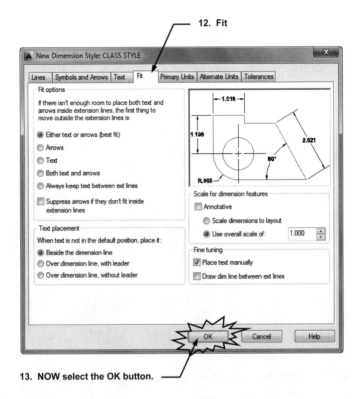

13. **NOW select the OK button.**

CREATING A NEW DIMENSION STYLE....continued

Your new style **"Class Style"** should be listed.

14. Select the **"Set Current"** button to make your new style "Class Style" the style that will be used.

New style listed here

14

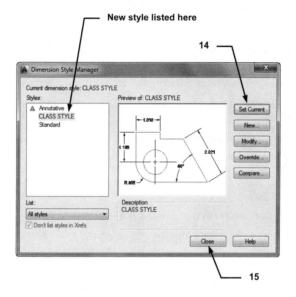

15

15. Select the **Close** button.

16. **Important:** Save your drawing again.

Note: You have successfully created a new "<u>Dimension Style</u>" called **"Class Style"**.
This style will be saved <u>in your drawing</u> after you save the drawing.
It is important that you understand that this dimension style resides <u>only</u> in this
drawing. If you open another drawing, this dimension style will not be there.

CREATING A DIMENSION SUB-STYLE

Previously you learned how to create a Dimension Style named Class Style. All of the dimensions created with that style appear identical because they have the same settings. Now you are going to learn how to create a **"Sub-Style"** of the Class Style.

For example:
If you wanted all of the **Diameter** dimensions to have a centerline automatically displayed but you did not want a centerline displayed when using the **Radius** dimension command. To achieve this, you must create a **"sub-style"** for all **Radius** dimensions.

Sub-styles have also been called "children" of the "Parent" dimension style.
As a result, they form a family.

A Sub-style is permanent, unlike the Override command, which is temporary.

Note: This sounds much more complicated than it is. Just follow the steps below.
It is very easy.

How to create a sub-style for Radius dimensions.

You will set the center mark to **None** for the Radius command only.
The Diameter command center mark will not change.

1. Open your drawing.

2. Select the **DIMENSION / STYLE** command. (Refer to page 3-5)

3. Select **"Class Style"** from the Style List.

4. Select the **NEW** button.

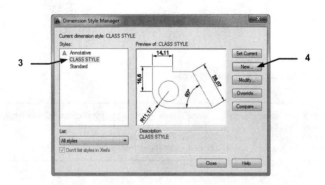

Continued on the next page...

CREATING A DIMENSION SUB-STYLE....continued

5. Change the "Use for" to: **Radius Dimensions**

6. Select **Continue**

This will turn gray. That's OK

6

5

7. Select the **Symbols and Arrows** tab.

8. Change the "Center Marks" to **NONE**

9. Select the **OK** button

7

8

9

10. You now have a sub-style that will <u>automatically override</u> the basic "Class Style" whenever you use a Radius command. Diameter dimensions will have centermarks and Radius dimensions will not.

Note: If you would like to keep this sub-style re-save the drawing or template.

10

EDITING DIMENSION TEXT VALUES

Sometimes you need to modify the dimension text value. You may add a symbol, note, or even change the text of an existing dimension. The following describes 2 methods.

Example: Add the word **"Max."** to the existing dimension value text.

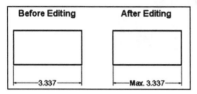

Method 1 (Properties Palette).

1. **Quick Properties** button must be **ON**.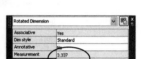
2. Select the dimension you want to override.
3. Scroll down to **Text override**

(Notice the actual <u>measurement</u> is directly above it.)

4. Type the new text: **Max** and **< >** and press **<enter>**
   (**< >** represents the associative text value 3.337)

Important: If you do not use < > the dimension will no longer be Associative.

Method 2 (Text Edit).

1. Type on the Command line: **ed <enter>** (This is the **text edit** command)
2. Select the dimension you want to edit.

Associative Dimension
If the dimension is <u>Associative</u> the dimension text will appear highlighted.

 **Before**

You may add text in front or behind the dimension text and it will remain Associative. Be careful not to disturb the dimension value text.

 **After**

Non Associative or Exploded Dimension
If the dimension value has been changed or exploded it will appear with a gray background and is not Associative.

3. Make the change.
4. Select the **OK** button.

EDITING THE DIMENSION POSITION

Sometimes dimensions are too close and you would like to stagger the text or you need to move an entire dimension to a new location. You can achieve this and more, using **"Grips"**.

Grips are great tools for repositioning dimensions.
Grips are small, solid-filled squares that are displayed at strategic points on objects.
You can drag these grips to stretch, move, rotate, scale, or mirror objects quickly.
Grips may be turned off by typing: **grips <enter>** then **0 <enter>** .

HOW TO USE GRIPS

1. Select the dimension to change. (no command can be in use while using grips)

2. Select one of the **blue** grips. It will turn to **"red"**. This indicates that grip is **"hot"**.
   The **"Hot"** grip is the **basepoint**.

3. Move the hot grip to the new location.

4. After editing you must press the **ESC** key to de-activate the grips.

The following is an example of how to use grips to quickly reposition dimensions.

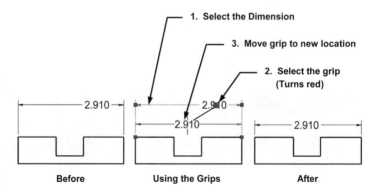

1. **Select the Dimension**

3. **Move grip to new location**

2. **Select the grip**
   (Turns red)

2.910   2.910   2.910

2.910

Before   Using the Grips   After

EDITING THE DIMENSION

ADDITIONAL EDITING OPTIONS USING THE SHORTCUT MENU

1. Select the dimension that you want to change.

2. Place the cursor on one of the grips shown below. (Do not press the mouse button)

3. Select an option from the short cut menu that appears.

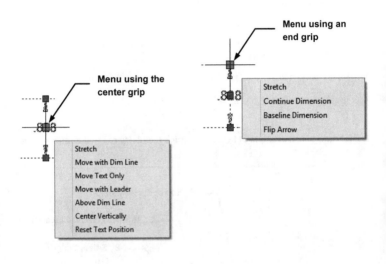

Menu using an end grip

Stretch
Continue Dimension
Baseline Dimension
Flip Arrow

Menu using the center grip

Stretch
Move with Dim Line
Move Text Only
Move with Leader
Above Dim Line
Center Vertically
Reset Text Position

MODIFY AN <u>ENTIRE</u> DIMENSION STYLE

After you have created a Dimension Style, you may find that you have changed your mind about some of the settings. You can easily change the entire Style by using the "Modify" button in the Dimension style Manager dialog box. This will not only change the Style for future use, but it will also <u>update dimensions already in the drawing</u>.

Note: if you do not want to update the dimensions already in the drawing, but want to make a change to a new dimension, refer to <u>Override</u>.

1. Select the Dimension Style Manager.

2. Select the Dimension Style that you wish to modify.

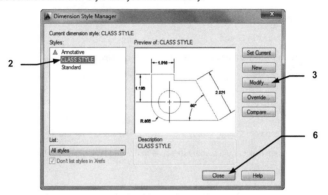

3. Select the **Modify button** from the Dimension Style Manager dialog box.

4. Make the desired changes to the settings.

5. Select the **OK** button.

6. Select the **Close** button.

Now look at your drawing. Have your dimensions updated?

Note:
The method above <u>will not</u> change dimensions that have previously been modified or exploded.

<u>Note: If some of the dimensions have not changed:</u>

1. Type: **-Dimstyle <enter>**   (notice the (-)dash in front of "dimstyle")

2. Type: **A <enter>**

3. Select dimensions to update and then **<enter>**

(Sometimes you have to give them a little nudge.)

OVERRIDE A DIMENSION STYLE

A dimension Override is a **temporary** change to the dimension settings.
An override **will not** affect existing dimensions. It will **only** affect **new dimensions**.
Use this option when you want a new dimension just a little bit different but you don't want to create a whole new dimension style and you don't want the existing dimensions to change either.

For example, if you want the new dimension to have a text height of .500 but you want the existing text to remain at .125 ht.

1. Select the **Dimension Style Manager**.

2. Select the **"Style"** you want to override. (Such as: Class Style)

3. Select the **Override** button.

4. Make the desired changes to the settings. (Such as: Text ht = .500)

5. Select the **OK** button.

6. Confirm the Override
   Look at the List of styles.
   Under the Style name, a sub heading of **<style overrides>** should be displayed.
   The description box should display the style name and the override settings.

7. The description box should display the style name and the override settings.

8. Select the **Close** button.

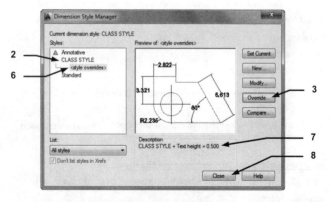

When you want to return to style **Class Style**, select **Class Style** from the styles list then select the **Set Current** button. Each time you select a different style, you must select the **Set Current** button to activate it.

EDIT AN INDIVIDUAL EXISTING DIMENSION

Sometimes you would like to modify the settings of an individual existing dimension. This can be achieved using the **Properties palette**.

1. Open the **Properties Palette**.

2. Select the dimension to change.

3. Select and change the desired settings.

Example: Change the dimension text height to .500

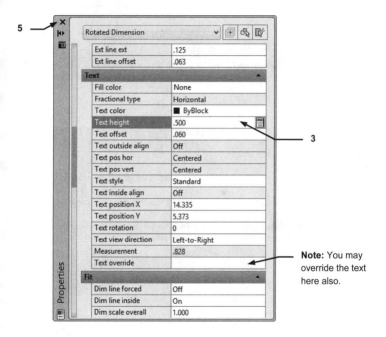

Rotated Dimension	
Ext line ext	.125
Ext line offset	.063
Text	
Fill color	None
Fractional type	Horizontal
Text color	ByBlock
Text height	.500
Text offset	.060
Text outside align	Off
Text pos hor	Centered
Text pos vert	Centered
Text style	Standard
Text inside align	Off
Text position X	14.335
Text position Y	5.373
Text rotation	0
Text view direction	Left-to-Right
Measurement	.828
Text override	
Fit	
Dim line forced	Off
Dim line inside	On
Dim scale overall	1.000

5 ←

3 ←

Note: You may override the text here also. ←

4. Press **<enter>** . *(The dimension should have changed)*

5. Close the **Properties Palette**.

6. Press the **<esc>** key to de-activate the grips.

<u>**Note:**</u> **The dimension will remain Associative.**

LINEAR DIMENSIONING

Linear dimensioning allows you to create horizontal and vertical dimensions.

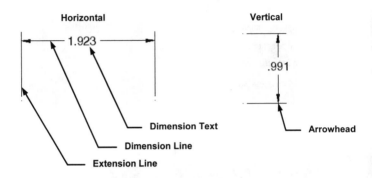

1. Select the **LINEAR** command using one of the following:

   Ribbon = Annotate Tab / Dimension Panel /
   or
   Keyboard = dimlinear <enter>

2. Specify first extension line origin or <select object>: ***snap to first extension line origin (P1)***.
3. Specify second extension line origin: ***snap to second extension line origin (P2)***.

4. Specify dimension line location or [Mtext/Text/Angle/Horizontal/Vertical/Rotated]: ***select where you want the dimension line placed (P3)***.

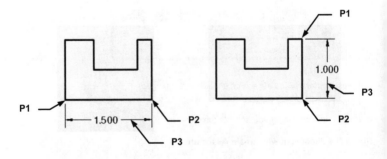

BASELINE DIMENSIONING

Baseline dimensioning allows you to establish a **baseline** for successive dimensions. The spacing between dimensions is automatic and should be set in the dimension style. (See 3-7, Baseline spacing setting)

A Baseline dimension must be used with an existing dimension. If you use Baseline dimensioning immediately after a Linear dimension, you do not have to specify the baseline origin.

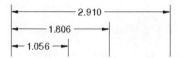

1. Create a <u>linear</u> dimension first (**P1** and **P2**).

2. Select the **BASELINE** command using one of the following:

Ribbon = Annotate Tab / Dimension Panel /
or
Keyboard = dimbaseline <enter>

3. Specify a second extension line origin or [Undo/Select] <Select>: *snap to the second extension line origin (P3)*.

4. Specify a second extension line origin or [Undo/Select] <Select>: *snap to P4*.

5. Specify a second extension line origin or [Undo/Select} <Select>: *select <enter> twice to stop*.

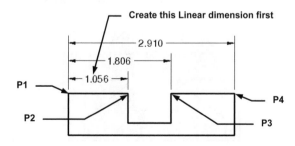

CONTINUE DIMENSIONING

Continue creates a series of dimensions in-line with an existing dimension. If you use the continue dimensioning immediately after a Linear dimension, you do not have to specify the continue extension origin.

1. Create a linear dimension first (**P1** and **P2** shown below)

2. Select the **CONTINUE** command using one of the following:

Ribbon = Annotate Tab / Dimension Panel /
or
Keyboard = dimcontinue <enter>

3. Specify a second extension line origin or [Undo/Select] <Select>: *snap to the second extension line origin (P3).*

4. Specify a second extension line origin or [Undo/Select] <Select>: *snap to the second extension line origin (P4).*

5. Specify a second extension line origin or [Undo/Select] <Select>: *press <enter> twice to stop.*

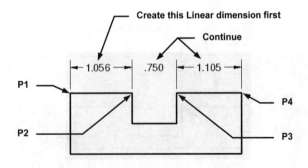

ALIGNED DIMENSIONING

The **ALIGNED** dimension command aligns the dimension with the angle of the object that you are dimensioning. The process is the same as Linear dimensioning. It requires two extension line origins and the placement of the dimension text location. (Example below)

1. Select the **ALIGNED** command using one of the following:

Ribbon = Annotate Tab / Dimension Panel /
or
Keyboard = dal <enter>

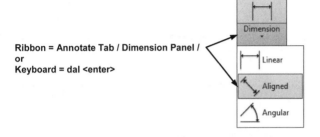

2. Specify first extension line origin or <select object>: *select the first extension line origin (P1)*

3. Specify second extension line origin: *select the second extension line origin (P2)*

4. Specify dimension line location or [Mtext/Text/Angle]: *place dimension text location (P3)*

   Dimension text = *the dimension value will appear here*

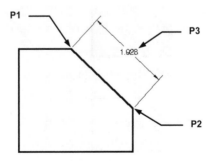

ANGULAR DIMENSIONING

The **ANGULAR** dimension command is used to create an angular dimension between two lines that form an angle. AutoCAD determines the angle between the selected lines and displays the dimension text followed by a degree (°) symbol.

1. Select the **ANGULAR** command using one of the following:

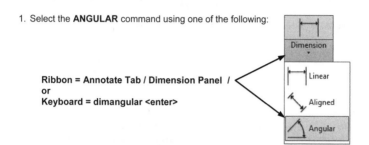

    Ribbon = Annotate Tab / Dimension Panel /
    or
    Keyboard = dimangular <enter>

2. Select arc, circle, line, or <specify vertex>: *select the first line that forms the*
        angle (P1) location is not important, <u>do not use object snap</u>.

3. Select second line: *select the second line that forms the angle (P2).*

4. Specify dimension arc line location or [Mtext/Text/Angle]: *place dimension text*
        Location (P3).

   Dimension text = *angle will be displayed here*

Any of the 4 angular dimensions shown below can be displayed by moving the cursor in the direction of the dimension after selecting the 2 lines (**P1** and **P2**) that form the angle.

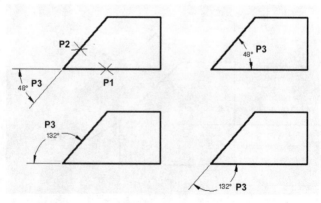

DIMENSIONING ARC LENGTHS

You may dimension the distance along an Arc. This is known as the **Arc length**.
Arc length is an <u>associative</u> dimension.

Example:

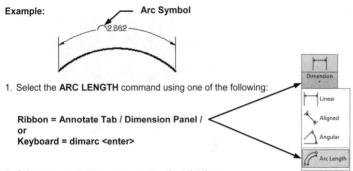

1. Select the **ARC LENGTH** command using one of the following:

   Ribbon = Annotate Tab / Dimension Panel /
   or
   Keyboard = dimarc <enter>

2. Select arc or polyline arc segment: ***select the Arc***

3. Specify arc length dimension location, or [Mtext/Text/Angle/Partial/Leader]:
   place the dimension line and text location

   Dimension text = ***dimension value will be shown here***

To differentiate the Arc length dimensions from Linear or Angular dimensions, arc
length dimensions display an arc (⌒) symbol by default. (Also called a "hat" or "cap")

The arc symbol may be displayed either <u>above</u>, or <u>preceding</u> the dimension text.
You may also choose not to display the arc symbol.

Example:

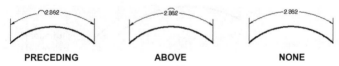

| PRECEDING | ABOVE | NONE |

Specify the placement of the arc symbol in the **Dimension Style / Symbols and
Arrows** tab or you may edit its position using the Properties Palette.

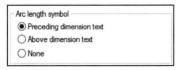

Continued on the next page...

DIMENSIONING ARC LENGTHS....continued

The extension lines of an Arc length dimension are displayed as <u>radial</u> if the included angle is <u>greater than 90 degrees</u>.

Example:

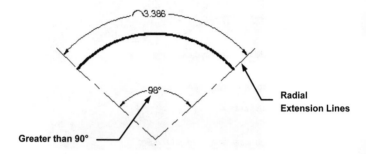

The extension lines of an Arc length dimension are displayed as <u>orthogonal</u> if the included angle is <u>less than 90 degrees</u>.

Example:

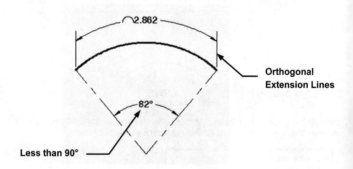

DIMENSIONING A LARGE CURVE

When dimensioning an arc the dimension line should pass through the center of the arc. However, for large curves, the true center of the arc could be very far away, even off the sheet. When the true center location cannot be displayed you can create a **"Jogged"** radius dimension.

Example:

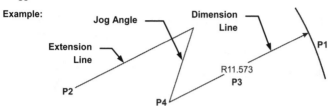

You can specify the jog angle in the **Dimension Style / Symbols and Arrows** tab.

1. Select the **JOGGED RADIUS** command using one of the following:

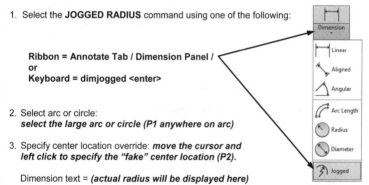

   Ribbon = Annotate Tab / Dimension Panel /
   or
   Keyboard = dimjogged <enter>

2. Select arc or circle:
   select the large arc or circle (P1 anywhere on arc)

3. Specify center location override: *move the cursor and left click to specify the "fake" center location (P2).*

   Dimension text = *(actual radius will be displayed here)*

4. Specify dimension line location or [Mtext/Text/Angle]: *move the cursor and left click to specify the location for the dimension text (P3).*

5. Specify jog location: *move the cursor and left click to specify the location for the jog (P4).*

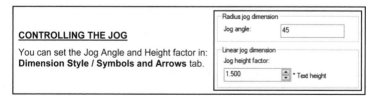

CONTROLLING THE JOG

You can set the Jog Angle and Height factor in: **Dimension Style / Symbols and Arrows** tab.

Radius jog dimension
Jog angle: 45

Linear jog dimension
Jog height factor:
1.500 * Text height

DIMENSIONING DIAMETERS

The **DIAMETER** dimensioning command should be used when dimensioning circles
and arcs of <u>more than 180 degrees</u>. AutoCAD measures the selected circle or arc and
displays the dimension text with the diameter symbol (Ø) in front of it.

1. Select the **DIAMETER** command using one of the following:

   Ribbon = Annotate Tab / Dimension Panel /
   or
   Keyboard = dimdiameter <enter>

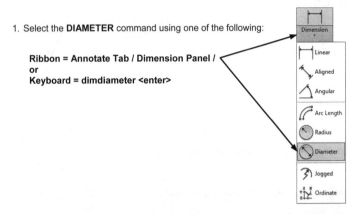

2. Select arc or circle: *select the arc or circle (P1) <u>do not use object snap</u>*.

   Dimension text = *the diameter will be displayed here.*

3. Specify dimension line location or [Mtext/Text/Angle]: *place dimension text
   location (P2)*

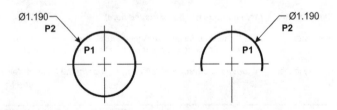

NOTE:

<u>Center marks</u> are automatically drawn as you use the diameter dimensioning
command. If the circle already has a center mark or you do not want a center mark, set
the center mark setting to **<u>NONE</u>** (**Dimension Style / Symbols and Arrows tab**)
before using Diameter dimensioning.

DIMENSIONING DIAMETERS....continued

Controlling the diameter dimension appearance.

If you would like your Diameter dimensions to appear as shown in the two examples below, you must change the **"Fit Options"** and **"Fine Tuning"** in the **Fit tab** in your Dimension Style.

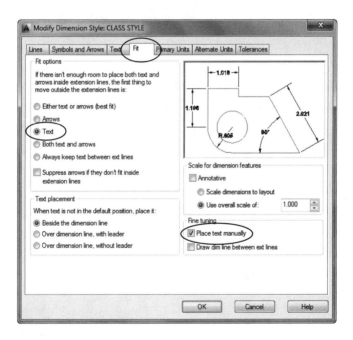

DIMENSIONING RADII

The **RADIUS** dimensioning command should be used when dimensioning arcs of less <u>than 180 degrees</u>. AutoCAD measures the selected arc and displays the dimension text with the radius symbol (**R**) in front of it.

1. Select the **RADIUS** command using one of the following:

   Ribbon = Annotate Tab / Dimension Panel /
   or
   Keyboard = dimradius <enter>

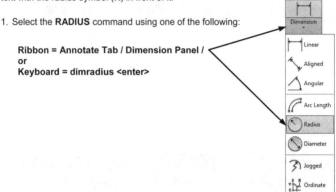

2. Select arc or circle: *select the arc (P1)* <u>do not use object snap</u>.

   Dimension text = *the radius will be displayed here*.

3. Specify dimension line location or [Mtext/Text/Angle]: *place dimension text location (P2)*

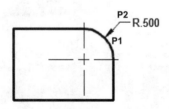

NOTE:

<u>Center marks</u> are automatically drawn as you use the radius dimensioning command. If you do not want a center mark, set the center mark setting to **<u>NONE</u>** (**Dimension Style / Symbols and Arrows tab**) before using Radius dimensioning.

DIMENSIONING RADII....continued

Controlling the radius dimension appearance.

If you would like your Radius dimensions to appear as shown in the example below, you must change the **"Fit Options"** and **"Fine Tuning"** in the **Fit tab** in your Dimension Style.

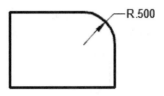

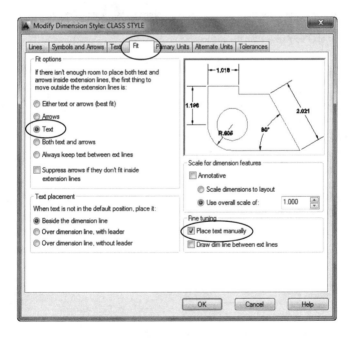

FLIP ARROW

You can easily flip the direction of the arrowhead using the **Flip Arrow** option.

How to Flip an arrowhead.

1. Select the dimension that you wish to Flip.

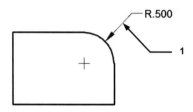

2. Rest your cursor on the arrow grip. (Do not press button, just rest cursor)

3. Select **Flip Arrow** from the shortcut menu.

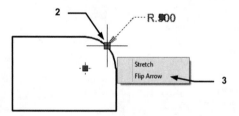

4. The arrowhead Flips.

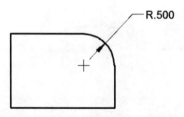

5. Press **Esc** and the grips will disappear. (Do not press <enter>).

QUICK DIMENSION

Quick Dimension creates multiple dimensions with one command. Quick Dimension can create Continuous, Staggered, Baseline, Ordinate, Radius and Diameter dimensions.

The following is an example of how to use the "Continuous" option. Staggered, Baseline, Diameter and Radius are explained on the following pages.

1. Select the **Quick Dimension** command using one of the following:

Ribbon = Annotate Tab / Dimension Panel /
or
Keyboard = qdim <enter>

Command: _qdim
Associative dimension priority = Endpoint
Select geometry to dimension

2. Select the objects to be dimensioned with a crossing window or pick each object

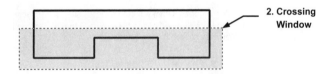

2. Crossing Window

3. Press **<enter>** to stop selecting objects.

4. Specify dimension line position, or
[Continuous/Staggered/Baseline/Ordinate/Radius/Diameter/datumPoint/Edit/settings] <Continuous>: *Select "C" <enter> for Continuous.*

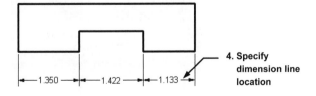

4. Specify dimension line location

QUICK DIMENSION....continued

STAGGERED

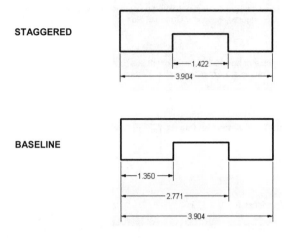

BASELINE

DIAMETER
1. You can use a crossing window. Qdim will automatically filter out any linear dims.
2. <u>Dimension Line Length</u> is determined by the **"Baseline Spacing"** setting in the Dimension Style. You may stretch it using grips.

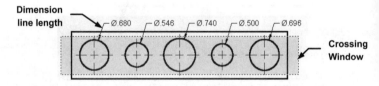

RADIUS
1. You can use a crossing window. Qdim will automatically filter out any linear dims.
2. Dimension Line Length is determined by the "Baseline Spacing" setting in the Dimension Style.

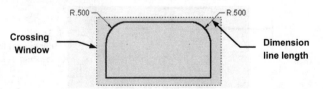

DIMENSION BREAKS

Occasionally extension lines overlap another extension line or even an object. If you do not like this you may use the **Dimbreak** command to break the intersecting lines. **Automatic** (described below) or **Manual** (described on next page) method may be used. You may use the **Remove** (described on page 3-37) option to remove the break.

<u>AUTOMATIC</u> DIMENSION BREAKS

To create an automatically placed dimension break, you select a dimension and then use the **Auto** option of the **DIMBREAK** command.

Automatic dimension breaks are updated any time the dimension or intersecting objects are modified.

You control the size of automatically placed dimension breaks on the <u>Symbols and Arrows</u> **tab of the Dimension Style dialog box.**

The specified size is affected by the dimension break size, dimension scale, and current annotation scale for the current viewport.

1. Select the **DIMBREAK** command using one of the following:

   Ribbon = Annotate Tab / Dimension Panel /
   or
   Keyboard = dimbreak <enter>

   Command: _Dimbreak
2. Select dimension to add / remove break or [Multiple]: *type M <enter>*
3. Select dimensions: *select dimension*
4. Select dimensions: *select another dimension*
5. Select dimensions: *select another dimension or <enter> to stop selecting*
6. Select object to break dimension or [Auto/Remove] <Auto>: *<enter>*

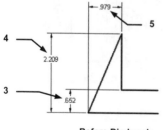

Before Dimbreak

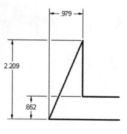

After Dimbreak

DIMENSION BREAK....continued

MANUAL DIMENSION BREAK

You can place a dimension break by <u>picking two points</u> on the dimension, extension, or leader line to determine the size and placement of the break.

Dimension breaks that are added manually by picking two points are not automatically updated if the dimension or intersecting object is modified. So if a dimension with a manually added dimension break is moved or the intersecting object is modified, you might have to restore the dimension and then add the dimension break again.

The size of a dimension break that is created by picking two points is not affected by the current dimension scale or annotation scale value for the current viewport.

1. Select the **DIMBREAK** command using one of the following:

Ribbon = Annotate Tab / Dimension Panel /
or
Keyboard = dimbreak <enter>

Command: _Dimbreak
2. Select dimension to add / remove break or [Multiple]: *select the dimension*

3. Select object to break dimension or [Auto/Manual/Remove] <Auto>: *type M <enter>*

4. Specify first break point: *select the first break point location* (Osnap should be off)

5. Specify second break point: *select the second break point location*

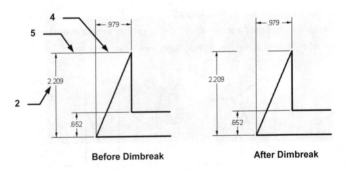

Before Dimbreak   **After Dimbreak**

The following objects can be used as cutting edges when adding a dimension break:
Dimension, Leader, Line, Circle, Arc, Spline, Ellipse, Polyline, Text, and Multiline text.

DIMENSION BREAK....continued

REMOVE THE BREAK

Removing the break is easy using the **<u>Remove</u>** option.

1. Select the **DIMBREAK** command using one of the following:

    Ribbon = Annotate Tab / Dimension Panel /
    or
    Keyboard = dimbreak <enter>

    Command: _Dimbreak
2. Select dimension to add / remove break or [Multiple]: *type M <enter>*

3. Select dimensions: *select a dimension*

4. Select dimensions: *select a dimension*

5. Select dimensions: *select another dimension or <enter> to stop selecting*

6. Select object to break dimension or [Auto/Remove] <Auto>: *type R <enter>*

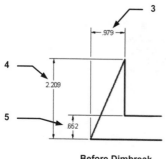

Before Dimbreak

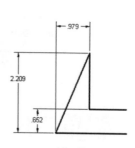

After Remove

JOG A DIMENSION LINE

Jog lines can be added to linear dimensions. Jog lines are used to represent a
dimension value that does not display the actual measurement.

Before you add a jog, the Jog angle and the height factor of the jog should be set in the
Symbols & Arrows tab within the Dimension Style Manager.

The height is calculated as a factor of the Text height.

For example:
if the text height was .250 and the jog height
factor was 1.500, the jog would be .375
Formula: (.250 ht. X 1.500 jog ht. factor= .375).

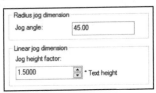

1. Select the **DIMJOGLINE** command using one of the following:

   Ribbon = Annotate Tab / Dimension Panel /
   or
   Keyboard = dimjogline <enter>

   Command: _DIMJOGLINE

2. Select dimension to add jog or [Remove]: **Select a dimension**

3. Specify jog location (or press ENTER): **Press <enter>**

4. After you have added the jog you can re-position it by using **Grips** and adjust the
   height of the jog symbol using the **Properties Palette**.

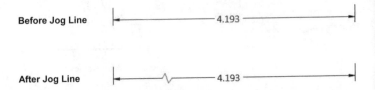

Before Jog Line

After Jog Line

REMOVE A JOG

1. Select the **DIMJOGLINE** command
   Command: _DIMJOGLINE

2. Select dimension to add jog or [Remove]: **type R <enter>**

3. Select jog to remove: **select the dimension**

ADJUST DISTANCE BETWEEN DIMENSIONS

The Adjust Space command allows you to adjust the distance between existing parallel linear and angular dimensions, so they are equally spaced. You may also align the dimensions to create a string.

1. Select the **ADJUST SPACE** command using one of the following:

   Ribbon = Annotate Tab / Dimension Panel /
   or
   Keyboard = dimspace <enter>

   Command: _DIMSPACE
2. Select base dimension: *Select the dimension that you want to use as the base dimension when equally spacing dimensions. (P1)*
3. Select dimensions to space: *select the next dimension to be spaced. (P2)*
4. Select dimensions to space: *select the next dimension to be spaced. (P3)*
5. Select dimensions to space: *continue selecting or press <enter> to stop.*
6. Enter value or [Auto] <Auto>: *enter a value or press <enter> for auto.*

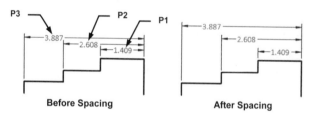

Before Spacing   After Spacing

Note: **Auto** creates a spacing value of twice the height of the dimension text.

For example: If the dimension text is 1/8", the spacing will be 1/4".

Aligning dimensions

Follow the steps shown above but when asked for the value enter **"0" <enter>**.

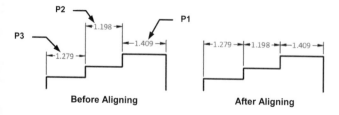

Before Aligning   After Aligning

MULTILEADER

A Multileader is a single object consisting of a **pointer**, **line**, **landing** and **content**. You may draw a Multileader, pointer first, tail first or content first.

To select the Multileader use:
Annotate tab / Leaders panel

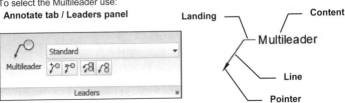

Create a single Multileader

1. Select the **Multileader** tool

   Command: _mleader
2. Specify leader arrowhead location or [leader Landing first/Content first/Options] <Options>: *specify arrowhead location (P1)*
3. Specify leader landing location: *specify the landing location (P2)*
4. The Multitext editor appears. *Enter text then select "Close Text Editor"*

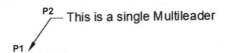

Add leader lines to existing Multileader

1. Select the **Add Leader** tool
2. Select a multileader: *click on the existing multileader line (P1)*
3. Specify leader arrowhead location: *specify arrowhead location (P2)*
4. Specify leader arrowhead location: *specify next arrowhead location (P3)*
5. Specify leader arrowhead location: *specify next arrowhead location (P4)*
6. Specify leader arrowhead location: *press <enter> to stop*

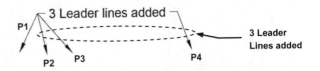

MULTILEADER....continued

<u>Remove a leader line from an existing Multileader</u>

1. Select the **Remove Leader** tool.

2. Select a multileader: *click anywhere on the existing multileader*
3. Specify leaders to remove: *select the leader line to remove*
4. Specify leaders to remove: *press <enter> to stop*

<u>Align Multileaders</u>

1. Select the **Align Multileader** tool

   Command: _mleaderalign
2. Select multileaders: *select the leaders to align (Note 1 and Note 3)*
3. Select multileaders: *press <enter> to stop selecting*
   Current mode: Use current spacing
4. Select multileader to align to or [Options]: *select the leader to align to (Note 2)*
5. Specify direction: *move the cursor and press left mouse button*

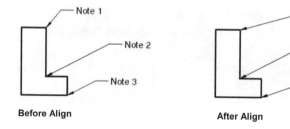

Note: <u>Collect Multileader</u> works best with Blocks and will be discussed
      in Section 4.

CREATE A MULTILEADER STYLE

You may create **Multileader styles** to control it's appearance.
This will be similar to creating a dimension style.

1. Open a drawing.

2. Select the **Multileader Style Manager** by selecting the ↘ on the Leader panel.

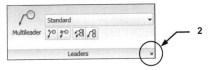

3. Select the **New** button.
4. Enter the New Style name: **Class ML style**
5. Select the **Continue** button.

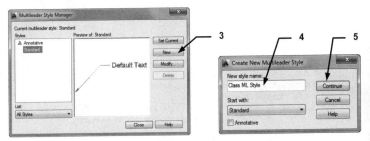

6. Select the **Leader Format** tab and change the settings as shown below.

<u>General</u> – You may set the
leader line to straight or spline
(curved). You may change the
color, linetype and lineweight.

<u>Arrowhead</u> – Select the symbol
for the pointer and size of
pointer.

<u>Leader break</u> - specify the size
of the gap in the leader line if
Dimension break is used.

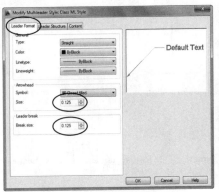

CREATE A MULTILEADER STYLE....continued

7. Select the **Leader Structure** tab and change the settings as shown below.

Constraints – Specify how many line segments to allow and on what angle.

Landing Settings – Specify the length of the Landing (the horizontal line next to the note) or turn it off by unchecking the "Automatically include landing" box.

Scale – "Annotation" will be discussed in Section 10.

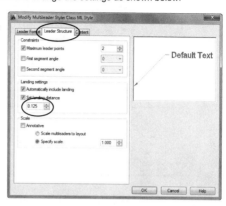

8. Select the **Content** tab and change the settings as shown below.

Multileader type – Select what to attach to the landing.

Text Options – Specify the leader note appearance.

Leader Connection – Specify how the note attaches to the landing.

9. Select the **OK** button.

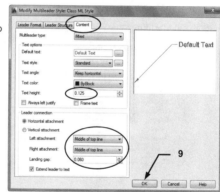

The **Class ML style** should be listed in the "Style" area.

10. Select **Set Current** button.

11. Select **Close** button.

12. **Save** your drawing again.

Now whenever you use this drawing the Multileader style **Class ML style** will be there.

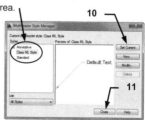

3-43

IGNORING HATCH OBJECTS

Occasionally, when you are dimensioning an object that has "Hatch Lines", your cursor will snap to the Hatch Line instead of the object that you want to dimension.
To prevent this from occurring, select the option **"IGNORE HATCH OBJECTS"**.

Example:

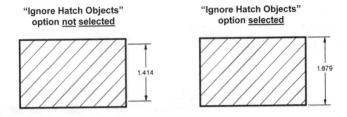

"Ignore Hatch Objects"
option <u>not</u> <u>selected</u>

1.414

"Ignore Hatch Objects"
option <u>selected</u>

1.679

How to select the "IGNORE HATCH OBJECTS" option

1. Type: **Options <enter>**

2. Select the **Drafting** tab.

3. Check the **Ignore Hatch Objects** box.

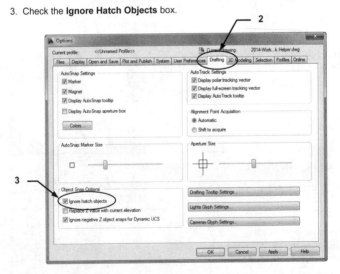

ORDINATE Dimensioning

Ordinate dimensioning is primarily used by the sheet metal industry. But many others are realizing the speed and tidiness this dimensioning process allows.

Ordinate dimensioning is used when the X and the Y coordinates, from one location, are the only dimensions necessary. Usually the part has a uniform thickness, such as a flat plate with holes drilled into it. The dimensions to each feature, such as a hole, originate from one "datum" location. This is similar to "baseline" dimensioning. Ordinate dimensions have only one datum. The datum location is usually the lower left corner of the object.

Ordinate dimensions appearance is also different. Each dimension has only one leader line and a numerical value. Ordinate dimensions do not have extension lines or arrows.

Example of Ordinate dimensioning:

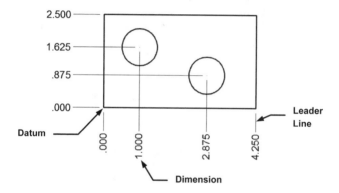

Note:
Ordinate dimensions can be Associative and are Trans-spatial. Which means that you can dimension in paperspace and the ordinate dimensions will remain associated to the object they dimension. (Except for Qdim ordinate)

Refer to the next page for step by step instructions to create Ordinate dimensions.

CREATING ORDINATE DIMENSIONS

1. Move the "Origin" to the desired "datum" location.
   Note: This must be done in Model Space.

2. Select the **ORDINATE** command using one of the following:

   Ribbon = Annotate Tab / Dimension Panel /
   or
   Keyboard = dimordinate <enter>

3. Select the first feature, using object snap.

4. Drag the leader line horizontally or vertically away from the feature.

5. Select the location of the "leader endpoint".
   (The dimension text will align with the leader line)

Use **"Ortho"** to keep the leader lines straight.

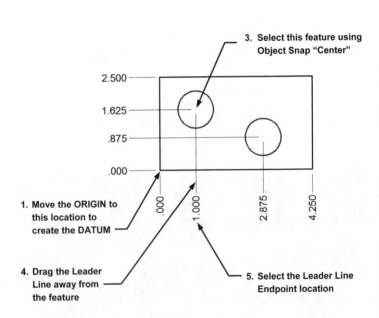

3. Select this feature using Object Snap "Center"

2.500

1.625

.875

.000

1. Move the ORIGIN to this location to create the DATUM

4. Drag the Leader Line away from the feature

5. Select the Leader Line Endpoint location

.000   1.000   2.875   4.250

JOG AN ORDINATE DIMENSION

If there is insufficient room for a dimension you may want to jog the dimension.
To **"jog"** the dimension, as shown below, turn **"Ortho" off** before placing the Leader
Line endpoint location. The leader line will automatically jog. With Ortho off, you can
only indicate the feature location and the leader line endpoint location, the leader line
will jog the way it wants to.

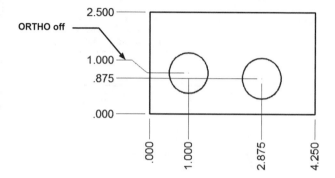

QUICK DIMENSION WITH ORDINATE DIMENSIONING

(Not available in version LT)

1. Select **DIMENSION / Quick Dimension** (Refer to page 3-33)
2. Select the geometry to dimension then press **<enter>**
3. Type **"O"** **<enter>** to select Ordinate
4. Type **"P"** **<enter>** to select the **datumPoint** option.
5. Select the datum location on the object. (use Object snap)
6. Drag the dimensions to the desired distance away from the object.

> **Note:** Qdim can be associative but is not trans-spatial.
> If the object is in Model Space, you must dimension in Model Space.

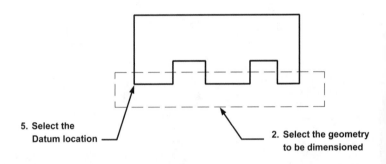

5. Select the
Datum location ———

2. Select the geometry
to be dimensioned

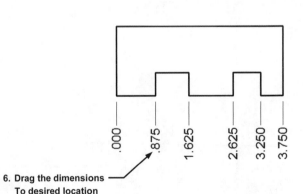

6. Drag the dimensions ———
To desired location

ALTERNATE UNITS

The options in this tab allow you to display inches as the primary units and the millimeter equivalent as alternate units. The millimeter value will be displayed inside brackets immediately following the inch dimension. Example: 1.225" [31.115mm]

1. Select a **DIMENSION STYLE** then the **Modify** button.

2. Select the **ALTERNATE UNITS** tab.

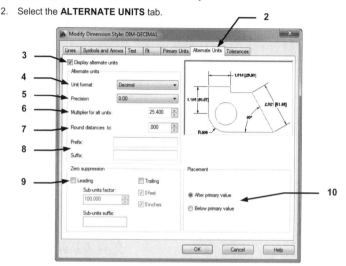

3. **Display alternate units**. Check this box to turn **ON** alternate units.

4. **Unit format**. Select the **Units** for the alternate units.

5. **Precision**. Select the Precision of the alternate units. This is independent of the Primary Units.

6. **Multiplier for all units**. The primary units will be multiplied by this number to display the alternate unit value.

7. **Round distance to**. Enter the desired increment to round off the alternate units value.

8. **Prefix / Suffix**. This allows you to include a Prefix or Suffix to the alternate units. Such as: type **mm** to the Suffix box to display **mm** (for millimeters) after the alternate units.

9. **Zero Suppression**. If you check one or both of these boxes, it means that the zero will not be drawn. It will be suppressed.

10. **Placement**. Select the desired placement of the alternate units. Do you want them to follow immediately after the Primary units or do you want the Alternate units to be below the primary units?

TOLERANCES

When you design and dimension a widget, it would be nice if when that widget was made, all of the dimensions were exactly as you had asked. But in reality this is very difficult and or expensive. So you have to decide what actual dimensions you could live with. Could the widget be just a little bit bigger or smaller and still work? This is why tolerances are used.

A **Tolerance** is a way to communicate, to the person making the widget, how much larger or smaller this widget can be and still be acceptable. In other words each dimension can be given a maximum and minimum size. But the widget must stay within that **"tolerance"** to be correct. For example: a hole that is dimensioned 1.00 +.06 -.00 means the hole is nominally 1.00 but it can be as large as 1.06 but can not be smaller than 1.00.

1. Select **DIMENSION STYLE / MODIFY**.

2. Select the **TOLERANCES UNITS** tab.

> **Note:** If the dimensions in the display look strange,
> Make sure **"Alternate Units"** are turned **OFF.**

3. **Method**
The options allows you to select how you would like the tolerances displayed. There are 5 methods: **None, Symmetrical, Deviation, Limits,** and **Basic.** (Basic is used in geometric tolerancing and will not be discussed at this time).

Refer to the next page for descriptions of the methods.

4. **Scaling the height**. This controls the height of the tolerance text. The entered value is a percentage of the primary text height. If .50 is entered, the tolerance text height will be 50% of the primary text height.

5. **Vertical position**. This controls the placement of the tolerance text in relation to the primary text. The options are Top, Middle and Bottom. Whichever option you select, it will align the tolerance text with the bottom of the primary text.

Continued on the next page...

TOLERANCES....continued

SYMMETRICAL is an equal bilateral tolerance. It can vary as much in the plus as in the negative. Because it is equal in the plus and minus direction, only the "Upper value" box is used. The "Lower value" box is grayed out.

Example of a Symmetrical tolerance:

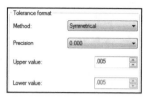

DEVIATION is an unequal bilateral tolerance. The variation in size can be different in both the plus and minus directions. Because it is different in the plus and the minus the "Upper" and "Lower" value boxes can be used.

Example of a Deviation tolerance:

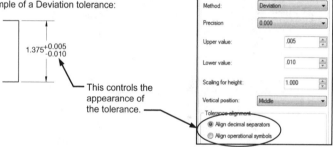

This controls the appearance of the tolerance.

Note: If you set the upper and lower values the same, the tolerance will be displayed as symmetrical.

LIMITS is the same as deviation except in how the tolerance is displayed. Limits calculates the plus and minus by adding and subtracting the tolerances from the nominal dimension and displays the results. Some companies prefer this method because no math is necessary when making the widget. Both "Upper and Lower" value boxes can be used.

Note: The "Scaling for height" should be set to "1".

Example of a Limits tolerance:

GEOMETRIC TOLERANCING

Geometric tolerancing is a general term that refers to tolerances used to control the form, profile, orientation, runout, and location of features on an object. Geometric tolerancing is primarily used for mechanical design and manufacturing. The instructions below will cover the Tolerance command for creating geometric tolerancing symbols and feature control frames.

1. Select the **TOLERANCE** command using one of the following:

   Ribbon = Annotate Tab / Dimension Panel ▼ /
   or
   Keyboard = tol <enter>

   The Geometric Tolerance dialog box, shown below, should appear.

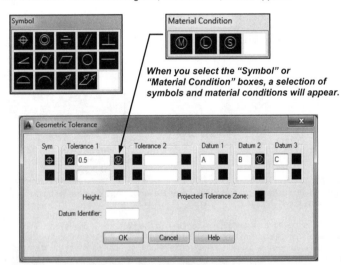

When you select the "Symbol" or "Material Condition" boxes, a selection of symbols and material conditions will appear.

2. Make your selections and fill in the tolerance and datum boxes.
3. Select the **OK** box.
4. The tolerance should appear attached to your cursor. Move the cursor to the desired location and press the left mouse button.

<u>Note</u>: *the size of the Feature Control Frame above, is determined by the height of the dimension text.*

GEOMETRIC TOLERANCES AND QLEADER

The **Qleader** command allows you to draw leader lines and access the dialog boxes used to create feature control frames in one operation. **Do not use Multileader**

1. Type **qleader <enter>**
2. Select **"Settings"**
3. Select the **Annotation** tab
4. Select the **Tolerance** option
5. Select **OK** button
6. Place the first leader point **P1**
7. Place the next point **P2**

8. Press **<enter>** to stop placing leader lines.

The Geometric Tolerance dialog box will appear.

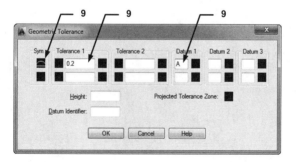

9. Make your selections and fill in the tolerance and datum boxes.
10. Select the **OK** button.

DATUM FEATURE SYMBOL

A datum in a drawing is identified by a **"datum feature symbol"**.

To create a *datum feature symbol:*

1. Select **Annotate** Tab / **Dimension Panel** ▼/
2. Type the **"datum reference letter"** in the **"Datum Identifier"** box.
3. Select the **OK** button.

Note: the size is determined by the text height setting within the current "dimension style".

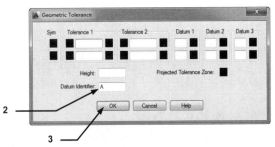

To create a *datum feature symbol* combined with a *feature control frame*:

1. Select **Annotate** Tab / **Dimension Panel** ▼/
2. Make your selections and fill in the tolerance.
3. Type the **"datum reference letter"** in the **"Datum Identifier"** box.
4. Select the **OK** button.

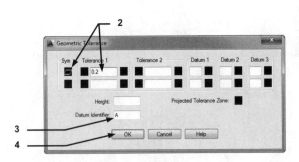

DATUM TRIANGLE

A datum feature symbol, in accordance with ASME Y14.5-2009, includes a
leader line and a datum triangle filled. You can create a block or you can use
the two step method below using Dimension / Tolerance and Leader.

1. Select **Dimension Style** and then **Override**

2. Change the Leader Arrow to **Datum triangle filled**

3. Select **Set Current** and **close**.

4. Type **qleader <enter>**

5. Place the 1st point (the triangle endpoint)

6. Place the 2nd point (Ortho ON)

7. Press **<enter> twice.**

8. Select **None** from the options.

If you were successful,
a datum triangle filled with
a leader line should appear.

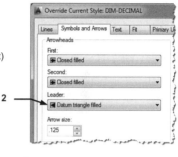

9. Next create a **datum feature symbol**.
   (Follow the instructions on the previous page.)

10. Now move the datum feature symbol to the endpoint of the leader line to create the
    symbol below left.

You are probably wondering why we didn't just type "A"
in the identifier box. That method will work if your leader
line is horizontal. But if the leader line is vertical, as
shown on the left, it will not work. (The example on the
right illustrates how it would appear)

TYPING GEOMETRIC SYMBOLS

If you want geometric symbols in the notes that you place on the drawing, you can
easily accomplish this using a font named **GDT.SHX**. This font will allow you to type
normal letters and geometric symbols, in the same sentence, by merely pressing the
SHIFT key when you want a symbol. <u>**Note:**</u> the **CAPS** lock must be **ON**.

1. First you must create a new text style using the **GDT.SHX** font.

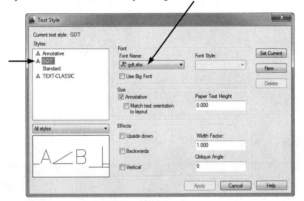

2. **CAPS LOCK** must be **ON**.

3. Select **Single Line** or **Multiline text**.

4. Now type the sentence shown below. When you want to type a symbol, press the
   SHIFT key and type the letter that corresponds to the symbol. For example: If you
   want the **diameter** symbol, press the **SHIFT** key and the **"N"** key. (Refer to the
   alphabet of letters and symbols shown above.)

$$3X \quad \varnothing.44 \quad \underline{\sqcup}\varnothing 1.06 \quad \overline{\vee}.06$$

Can you decipher what it says?

(Drill (3) .44 diameter holes with a 1.06 counterbore diameter .06 deep)

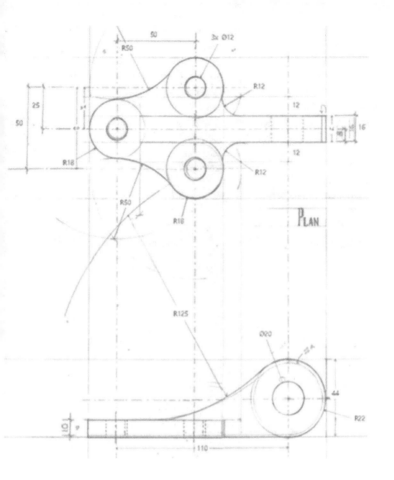

Section 4
Drawing Entities

ARC

There are 10 ways to draw an **ARC** in AutoCAD. Not all of the ARCS options are easy to create so you may find it is often easier to **trim a Circle** or use the **Fillet** command.

On the job, you will probably only use 2 of these methods. Which 2 depends on the application.

An **ARC** is a segment of a circle and must be less than 360 degrees.

By default, ARCS are drawn counter-clockwise. In some cases you can enter a negative input to draw in a clockwise direction.

1. Select the **ARC** Command using one of the following:

   Ribbon = Home Tab / Draw Panel /
   or
   Keyboard = A <enter>

2. Refer to the following pages for examples of each method listed below.

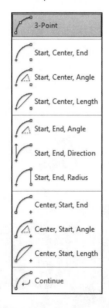

Note:
AutoCAD **2014** has an option that allows you to change the direction of ARCS. When creating an ARC hold down the **Ctrl** key to draw in a clockwise direction. This option is not available in the 2013 version.

3 POINT

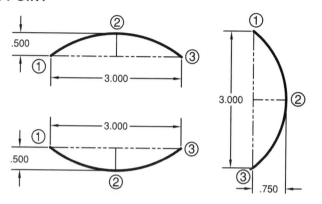

AutoCAD 2013/2014: You can draw this Arc in either a counter-clockwise or clockwise direction.

START, CENTER, END

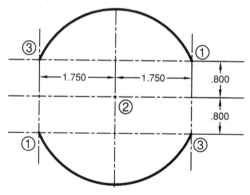

AutoCAD 2013: These Arcs are always drawn counter-clockwise.

AutoCAD 2014: By default these Arcs are drawn counter-clockwise. Hold down the **Ctrl** key to draw in a clockwise direction.

START, CENTER, ANGLE

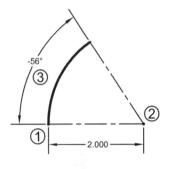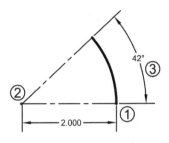

AutoCAD 2013/2014: Positive angles draw the Arc counter-clockwise, negative angles draw the Arc clockwise.

AutoCAD 2014: To reverse the directions hold down the **Ctrl** key.

START, CENTER, LENGTH

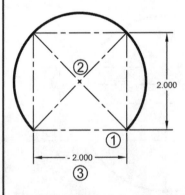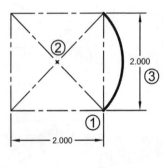

AutoCAD 2013/2014: Positive Chord length draws the small segment counter-clockwise, Negative Chord length draws the large segment counter-clockwise.

AutoCAD 2014: To reverse the directions hold down the **Ctrl** key.

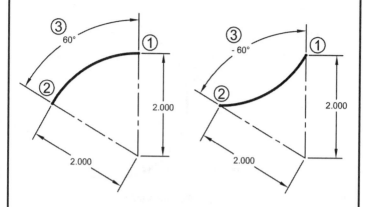

START, END, ANGLE

AutoCAD 2013/2014: Positive angles draw the Arc counter-clockwise, negative angles draw the Arc clockwise.

AutoCAD 2014: To reverse the directions hold down the **Ctrl** key.

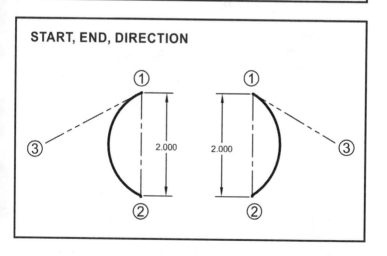

START, END, DIRECTION

START, END, RADIUS

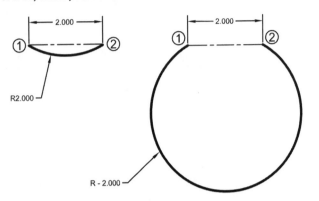

AutoCAD 2013/2014: Positive radius draws the small segment, negative radius draws the large segment.

AutoCAD 2014: To reverse the directions hold down the **Ctrl** key.

CENTER, START, END

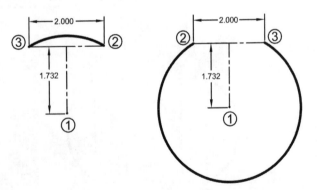

AutoCAD 2013/2014: Draws the Arc counter-clockwise.

AutoCAD 2014: To reverse the directions hold down the **Ctrl** key.

CENTER, START, ANGLE

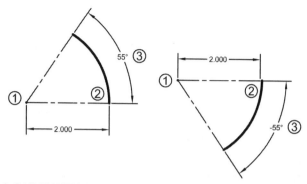

AutoCAD 2013/2014: Positive angles draw the Arc counter-clockwise, negative angles draw the Arc clockwise.

AutoCAD 2014: To reverse the directions hold down the **Ctrl** key.

CENTER, START, LENGTH

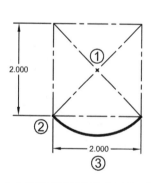 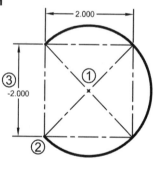

AutoCAD 2013/2014: Positive Chord length draws the small segment counter-clockwise, Negative Chord length draws the large segment counter-clockwise.

AutoCAD 2014: To reverse the directions hold down the **Ctrl** key.

CONTINUE

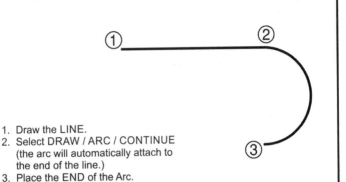

1. Draw the LINE.
2. Select DRAW / ARC / CONTINUE
   (the arc will automatically attach to
   the end of the line.)
3. Place the END of the Arc.

DRAWING A SPIRAL

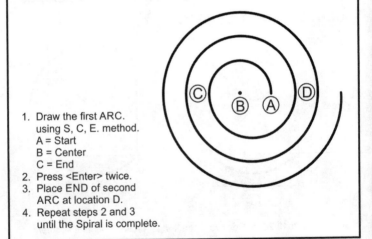

1. Draw the first ARC.
   using S, C, E. method.
   A = Start
   B = Center
   C = End
2. Press <Enter> twice.
3. Place END of second
   ARC at location D.
4. Repeat steps 2 and 3
   until the Spiral is complete.

BLOCKS

A BLOCK is a group of objects that have been converted into ONE object. A Symbol, such as a transistor, bathroom fixture, window, screw or tree, is a typical application for the block command. First a BLOCK must be created. Then it can be INSERTED into the drawing. An inserted Block uses less file space than a set of objects copied.

CREATING A BLOCK

1. First draw the objects that will be converted into a Block.

For this example a circle and 2 lines are drawn.

2. Select the **CREATE BLOCK** command using one of the following:

Ribbon = Insert Tab / Block Definition Panel /
or
Keyboard = B <enter>

3. Enter the New Block name in the **Name** box.

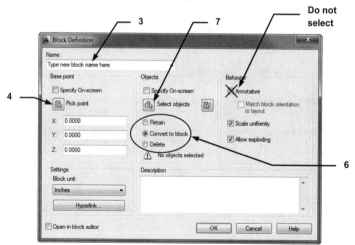

Continued on the next page...

4-9

BLOCKS....continued

4. Select the **Pick Point** button. (Or you may type the **X**, **Y** and **Z** coordinates.)
   The Block Definition box will disappear and you will return temporarily to the drawing.

5. Select the location where you would like the insertion point for the Block.
   Later when you insert this block, the block will appear on the screen attached to the cursor at this insertion point. Usually this point is the CENTER, MIDPOINT or ENDPOINT of an object.

Notice the coordinates for the base point are now displayed. (Don't worry about this. Use Pick Point and you will be fine

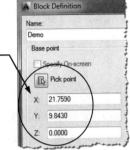

6. Select one of the options described below.

   It is important that you select one and understand the options below.

   Retain
   If this option is selected, the original objects will stay visible on the screen after the block has been created.

   Convert to block
   If this option is selected, the original objects will disappear after the block has been created, but will immediately reappear as a block. It happens so fast you won't even notice the original objects disappeared.

   Delete
   If this option is selected, the original objects will disappear from the screen after the block has been created. (This is the one I use most of the time)

7. Select the **Select Objects** button.

   The Block Definition box will disappear and you will return temporarily to the drawing.

8. Select the objects you want in the block, then press **<enter>**.

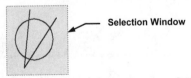

Selection Window

Continued on the next page...

BLOCKS....continued

The **Block Definition** box will reappear and the objects you selected should be displayed in the Preview Icon area.

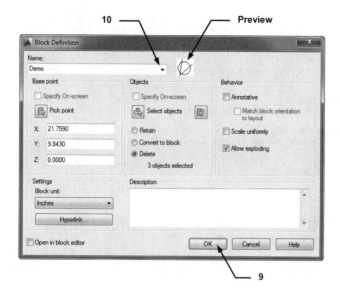

9. Select the **OK** button.
   The new block is now stored in the drawing's block definition table.

10. To verify the creation of this Block, select **BLOCK CREATE** again, and select the **Name** (▼). A list of all the blocks, in this drawing, will appear.

Continued on the next page...

BLOCKS....continued

ADDITIONAL DEFINITIONS OF OPTIONS

Block Units
You may define the units of measurement for the block. This option is used with the "Design Center" to drag and drop with Autoscaling. The Design Center is an advanced option and is not discussed in this book.

Hyperlink
Opens the **Insert Hyperlink** dialog box which you can use to associate a hyperlink with the block.

Description
You may enter a text description of the block.

Scale Uniformly
Specifies whether or not the block is prevented from being scaled non-uniformly during insertion.

Allow Exploding
Specifies whether or not the block can be exploded after insertion.

HOW LAYERS AFFECT BLOCKS

If a block is created on Layer 0:

1. When the block is inserted, it will take on the properties of the current layer.
2. The inserted block will reside on the layer that was current at the time of insertion.
3. If you Freeze or turn Off the layer the block was inserted onto, the block will disappear.
4. If the Block is **Exploded**, the objects included in the block will revert to their original properties of **layer 0**.

If a block is created on Specific layers:

1. When the block is inserted, it will retain its own properties. It **will not** take on the properties of the current layer.
2. The inserted block **will reside** on the current layer at the time of insertion.
3. If you **freeze** the layer that was current at the time of insertion the block will disappear.
4. If you turn **off** the layer that was current at the time of insertion the block will not disappear.
5. If you **freeze** or turn **off** the blocks original layers the block will disappear.
6. If the Block is **Exploded**, the objects included in the block will go back to their original layer.

INSERTING BLOCKS

A **BLOCK** can be inserted at any location within the drawing. When inserting a Block you can **SCALE** or **ROTATE** it.

1. Select the **INSERT** command using one of the following:

   Ribbon = Insert Tab / Block Definition Panel /
   or
   Keyboard = insert <enter>

   Insert

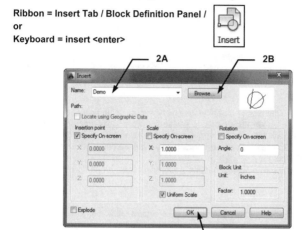

2. Select the **BLOCK** name.
   a. If the block is already in the drawing that is open on the screen, you may select the block from the drop down list shown above
   b. If you want to insert an entire drawing, select the Browse button to locate the drawing file.

3. Select the **OK** button.

   This returns you to the drawing and the selected block should be attached to the cursor.

4. Select the insertion location for the block by moving the cursor and pressing the left mouse button or typing coordinates.

   Command: _insert
   Specify insertion point or **[Basepoint/Scale/Rotate]:**

   NOTE: If you want to change the **Basepoint, Scale** or **Rotate** the block before you actually place the block, press the right hand mouse button and you may select an option from the menu or select an option from the command line menu shown above.

   You may also **"preset"** the insertion point, scale or rotation. This is discussed on the next page.

Continued on the next page...

INSERTING BLOCKS....continued

PRESETTING THE <u>INSERTION POINT</u>, <u>SCALE</u> or <u>ROTATION</u>

You may preset the **Insertion point, Scale** or **Rotation** in the <u>**INSERT**</u> box instead of at the command line.

1. Remove the check mark from any of the **"Specify On-screen"** boxes.
2. Fill in the appropriate information describe below:

Insertion point
Type the X and Y coordinates <u>from the Origin</u>. The Z is for 3D only.
The example below indicates the block's insertion location will be 5 inches in the X direction and 3 inches in the Y direction, <u>from the Origin</u>.

Scale
You may scale the block proportionately by typing the scale factor in the X box and then check the <u>Uniform Scale box</u>. If you selected the "Scale uniformly" box when creating the block this option is unnecessary and not available.
If the block is to be scaled non-proportionately, type the different scale factors in both X and Y boxes.
The example below indicates that the block will be scale proportionate at a factor of 2.

Rotation
Type the desired rotation angle relative to its current rotation angle.
The example below indicates the block will be rotated 45 degrees from its originally created angle orientation.

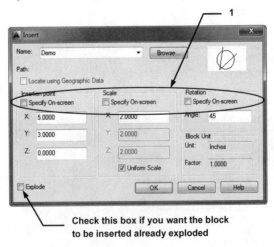

Check this box if you want the block
to be inserted already exploded

RE-DEFINING A BLOCK

How to change the design of a block previously inserted.

1. Select Block Editor using one of the following:

   Ribbon = Insert Tab / Block Definition Panel /
   or
   Keyboard = bedit <enter>

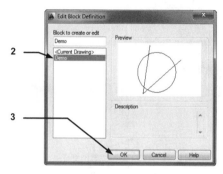

*The **"Edit Block Definition"** dialog box will appear.*

2. Select the name of the Block that you wish to change.

3. Select the **OK** button.

4. The Block that you selected should appear large on the screen.
   You may now make any additions or changes to the block. You can change tabs
   and use other panels such as Draw and Modify. But you must return to the **Block
   Editor** tab to complete the process.

5. Return to the **Block Editor** button if you selected any other tab while editing.

6. Select the **Save block** tool from the **Open/Save** panel.

7. Select the **Close Block Editor** tool.

You will be returned to the drawing and **all previously inserted** blocks with the **same
name** will be updated with the changes that you made.

PURGE UNWANTED AND UNUSED BLOCKS

You can remove a block reference from your drawing by erasing it; however, the block definition remains in the drawing's block definition table. To remove any **unused** blocks, dimension styles, text styles, layers and linetypes you may use the **PURGE** command.

How to delete unwanted and unused blocks from the current drawing.

1. Select the **PURGE** command using one of the following:

   Ribbon = None
   or
   Application Menu = Drawing Utilities / Purge
   or
   Keyboard = purge <enter>

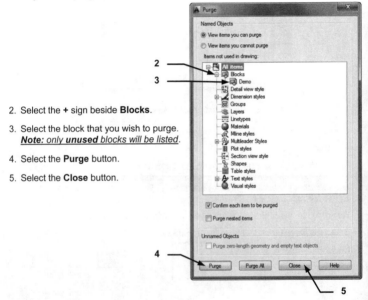

2. Select the **+** sign beside **Blocks**.

3. Select the block that you wish to purge.
   *Note: only **unused** blocks will be listed*.

4. Select the **Purge** button.

5. Select the **Close** button.

WHERE ARE BLOCKS SAVED?

When you create a Block it is saved **within the drawing you created it in**.

(If you open another drawing you will not find that block.)

MULTILEADER AND BLOCKS

In Section 3 you learned about Multileaders and how easy and helpful they are to use. Now you will learn another user option within the **Multileader Style Manager** that allows you to <u>attach a **pre-designed Block**</u> to the landing end of the Leader. To accomplish this, you must <u>first create the style</u> and <u>then you may use it</u>.

Here are a few examples of multileaders with pre-designed blocks attached:

STEP 1. CREATE A NEW MULTILEADER STYLE

1. Select the **Multileader Style Manager** tool using:

   Ribbon = Annotate tab / Leaders panel / ⅃

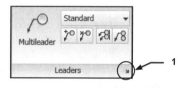

2. Select the **New** button.

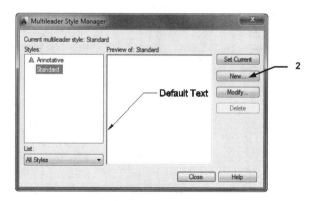

Continued on the next page...

MULTILEADER AND BLOCKS....continued

3. Enter **New Style Name**.

4. Select a style to **Start with:**

5. Select **Annotative** box.

6. Select **Continue** button.

7. Select the **Content** tab.

8. **Multileader Type:** Select Block from drop down menu.

9. **Source Block:** Select **Circle** from the drop down menu.
   Note: you have many choices here. These are AutoCAD pre-designed blocks <u>with Attributes</u>.

10. **Attachment:** Select **Center Extents**.
    Note: this selection works best with Circle but you will be given <u>different choices</u> depending on which <u>Source block you select</u>.

11. **Color:** Select **Bylayer**
    Bylayer works best. It means, it acquires the **color** of the current layer setting.

12. **Scale:** Select 1.000

13. Select **OK** button.

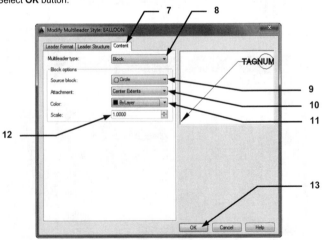

Continued on the next page...

MULTILEADER AND BLOCKS....continued

Your new multileader style should be displayed in the Styles list.

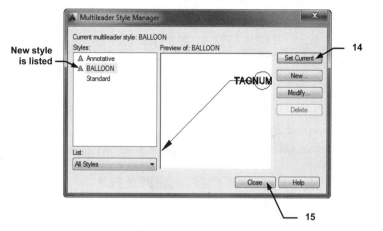

14. Select **Set Current**.

15. Select the **Close** button.

MULTILEADER AND BLOCKS....continued

STEP 2. USING THE MULITLEADER WITH A BLOCK STYLE

1. Select the **Annotate tab / Leaders panel**.

2. Select the **Style** from the Multileader drop down list.

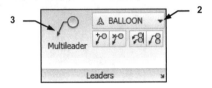

3. Select the **Multileader** tool.

4. The **"Select Annotation Scale"** box may appear. Select **OK** for now.

5. Specify leader arrowhead location or [leader Landing first/Content first/Options] <Options>: *place the desired location of the arrowhead*

6. Specify leader landing location: *place the desired location of the landing*

> *The next step is where the pre-assigned "Attributes" activate.*
> *Refer to the Advanced Workbook for Attributes.*

7. Enter attribute values
   Enter tag number <TAGNUMBER>: *type number or letter <enter>*

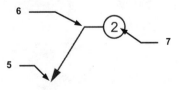

COLLECT MULTILEADER

In Section 3 you learned how to ADD, REMOVE and ALIGN multileaders. Now you will learn how to use the **COLLECT** multileader tool.

If you have multiple Leaders pointing to the same location or object you may wish to **COLLECT** them into one Leader.

Example:

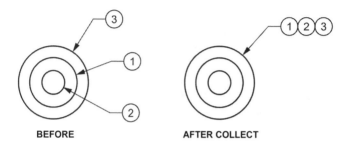

BEFORE   AFTER COLLECT

HOW TO USE COLLECT MULTILEADER

1. Select the **Collect Mutileader** tool.

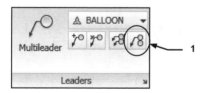

2. Select the Multileaders that you wish to combine then press **<enter>**.
   Note: Select them one at a time <u>in the order you wish them to display</u>.

   Such as (1, 2, 3, A, B, C, etc)

3. Place the combined Leader location. (**Ortho** should be **OFF**)

CENTERMARK

CENTERMARKS can ONLY be drawn with circular objects like Circles and Arcs. You set the size and type.

The Center Mark has three <u>types</u>, **None**, **Mark** and **Line** as shown below.

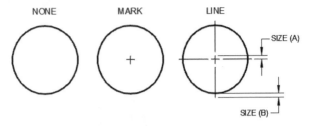

NONE   MARK   LINE

SIZE (A)

SIZE (B)

What does "SIZE" mean?

The size setting determines both, (A) the length of half of the intersection line and (B) the length extending beyond the circle. (See above right)

Where do you set the CENTERMARK "TYPE" and "SIZE"

1. Select the **Dimension Style** command.
2. Select: **Modify** or **Override**.
3. Select: **Symbols** and **Arrows** tab.
4. Select the **Center mark type**.
5. Set the **Size**.

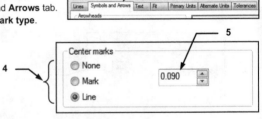

To create a CENTER MARK

1. Select the **CENTERMARK** command using one of the following:

   Ribbon = Annotate Tab / Dimension Panel / ▼
   or
   Keyboard = dce <enter>

2. Select arc or circle: *select the arc or circle with the cursor*.

CIRCLE

There are 6 options to create a circle.

The default option is **"Center, radius"**. (Probably because that is the most common method of creating a circle.)

We will try the **"Center, radius"** option first.

1. Start the **CIRCLE** command by using one of the following:

   Ribbon = Home tab / Draw Panel /
   or
   Keyboard = C <enter>

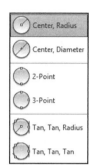

2. The following will appear on the command line:
   Command: _circle Specify center point for circle or [3P/2P/Ttr (tan tan radius)]:

3. Locate the center point for the circle by moving the cursor to the desired location in the drawing area (**P1**) and press the left mouse button.

4. Now move the cursor away from the center point and you should see a circle forming.

5. When the circle is the size desired (**P2**), press the left mouse button, or type the radius and then press **<enter>**.

Note: To use one of the other methods described below, first select the Circle command, then select one of the other Circle options.

Center, Radius: (Default option)

1. Specify the center (**P1**) location.

2. Specify the Radius (**P2**).
   (Define the Radius by moving the cursor or typing radius)

Center, Diameter:

1. Specify the center (**P1**) location.

2. Specify the Diameter (**P2**). (Define the Diameter by moving the cursor or typing the Diameter)

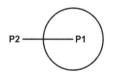

Continued on the next page...

CIRCLE....continued

2 Points:

1. Select the 2 point option.
2. Specify the 2 points (**P1** and **P2**) that will determine the Diameter.

3 Points:

1. Select the 3 Point option.
2. Specify the 3 points (**P1**, **P2** and **P3**) on the circumference. The Circle will pass through all three points.

Tangent, Tangent, Radius:

1. Select the Tangent, Tangent, Radius option.
2. Select two objects (**P1** and **P2**) for the Circle to be tangent to by placing the cursor on the object and pressing the left mouse button.
3. Specify the radius.

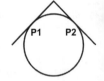

Tangent, Tangent, Tangent:

1. Select the Tangent, Tangent, Tangent option.
2. Specify three objects (**P1**, **P2** and **P3**) for the Circle to be tangent to by placing the cursor on each of the objects and pressing the left mouse button. (AutoCAD will calculate the diameter.)

DONUT

A Donut is a circle with *width*. You will define the **Inside** and **Outside** diameters.

Outside Diameter   Inside Diameter

1. Select the **DONUT** command using one of the following:

   Ribbon = Home Tab / Draw Panel ▼ /
   or
   Keyboard = DO <enter>

2. The following prompts will appear on the command line:

   Command: _donut
   Specify inside diameter of donut: *type the inside diameter <enter>*
   Specify outside diameter of donut: *type the outside diameter <enter>*
   Specify center of donut or <exit>: *place the center of the first donut*
   Specify center of donut or <exit>: *place the center of the second donut or*
   enter> to stop

Note:
It will continue to create more donuts until you press **<enter>** to stop the command.

Controlling the "FILL MODE"

1. Command: *type FILL <enter>*

2. Enter mode [ON / OFF] <OFF>: *type ON or OFF <enter>*

3. Type *REGEN <enter>* to regenerate the drawing to show the latest setting of the
   FILL mode.

FILL = ON

FILL = OFF

ELLIPSE

There are 3 methods to draw an Ellipse. You may (1) specify 3 points of the axes, (2) define the center point and the axis points or (3) define an elliptical Arc.

The following 3 pages illustrates each of the methods.

AXIS END METHOD

1. Select the **ELLIPSE** command using one of the following:

Ribbon = Home Tab / Draw Panel /
or
Keyboard = EL <enter>

2. The following prompts will appear on the command line:

Command: _ellipse
Specify axis endpoint of ellipse or [Arc/Center]: *place the first point of either the major or minor axis (P1)*.

Specify other endpoint of axis: *place the other point of the first axis (P2)*

Specify distance to other axis or [Rotation]: *place the point perpendicular to the first axis (P3)*.

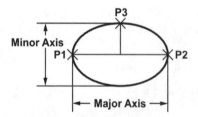

Specifying Major Axis first (P1/P2), then Minor Axis (P3)

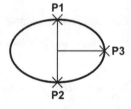

Specifying Minor Axis first (P1/P2), then Major Axis (P3)

Continued on the next page...

ELLIPSE....continued

CENTER METHOD

1. Select the **ELLIPSE** command using one of the following:

   Ribbon = Home Tab / Draw Panel /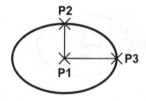
   or
   Keyboard = EL <enter> C <enter>

2. The following prompts will appear on the command line:

   Command: _ellipse

   Specify center of ellipse: *place center of ellipse (P1)*

   Specify endpoint of axis: *place first axis endpoint (either axis) (P2)*

   Specify distance to other axis or [Rotation]: *place the point perpendicular to the first axis (P3)*

Continued on the next page...

ELLIPSE....continued

ELLIPTICAL ARC METHOD

1. Select the **ELLIPSE** command using one of the following:

   Ribbon = Home Tab / Draw Panel / [icon] **/**   [Elliptical Arc]
   or
   Keyboard = EL <enter> A <enter>

2. The following prompts will appear on the command line:

   Command: _ellipse

   Specify axis endpoint of elliptical arc or [center]: *type C <enter>*

   Specify center of axis: *place the center of the elliptical arc (P1)*

   Specify endpoint of axis: *place first axis point (P2)*

   Specify distance to other axis or [Rotation]: *place the endpoint perpendicular to
   the first axis (P3)*

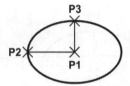

Specify start angle or [Parameter]: *place the start angle (P4)*

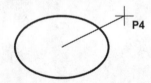

Specify end angle or [Parameter/Included angle]: *place end angle (P5)*

HATCH

The **HATCH** command is used to create hatch lines for section views or filling areas with specific patterns.

To draw **hatch** you must start with a closed boundary. A closed boundary is an area completely enclosed by objects. A rectangle would be a closed boundary. You simply place the cursor inside the closed boundary or select objects.

Note:
A Hatch set is one object.
It is good drawing management to always place Hatch on it's own layer.
Use Layer Hatch. You may also make Hatch appear or disappear with the **FILL** command.

HOW TO PLACE HATCH

1. Draw a Rectangle.

2. Select the **HATCH** command using one of the following:

Ribbon = Home Tab / Draw Panel /
or
Keyboard = BH <enter>

The **"Hatch Creation"** ribbon tab appears automatically.

5

3. Place the cursor inside the Rectangle (a closed boundary).

A hatch pattern preview will appear.

4. Press the **left mouse button** to accept the Hatch.

5. Select **Close Hatch Creation** or press **<enter>**

HATCH PROPERTIES

When you select the **Hatch** command the **Hatch Creation** ribbon tab appears automatically. The panels on this tab help set the properties of the Hatch. You should set the properties desired previous to placing the hatch set although you can easily edit an existing hatch set.

BOUNDARIES Panel

The Boundaries panel allows you to choose what method you will use to select the hatch boundary.

Pick Point:
Pick Point is the default selection. When you select the Hatch command AutoCAD assumes that you want to use the Pick Points method. You merely place the cursor in the closed area to select the boundary. The Hatch set preview will appear. Press the left mouse button to accept.

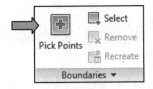

Select and Remove:
You may select or remove objects to a boundary.

Note: Remove will not be available unless you click on **Select**.

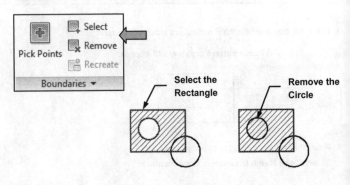

HATCH PROPERTIES....continued

PATTERN Panel

Select to display
additional patterns

The **PATTERN panel** displays the Hatch swatches that relate to the Hatch Type that
has been selected in the Properties panel. Refer to Properties Panel below.

PROPERTIES Panel

HATCH TYPE

When you select the drop down arrow ▼ you may select one of the Hatch types:
Solid, **Gradient**, **Pattern** or **User Defined**.

Note: Hatch Types will be explained in more detail on pages 4-35 through 4-38.

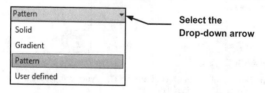

Select the
Drop-down arrow

When you select a Hatch Type from the drop down list, the Pattern panel displays the
related hatch swatches from which to select.

| Pattern | Solid | User Defined | Gradient |

Continued on the next page...

HATCH PROPERTIES....continued

PROPERTIES Panel....continued

HATCH COLOR

This color selection is specific to the Hatch and will not affect any other objects.

BACKGROUND COLOR

You may select a background color for the hatch area.

HATCH TRANSPARENCY

Displays the selected transparency setting. You may select to use the current setting for the drawing, the current layer setting or select a specific value.

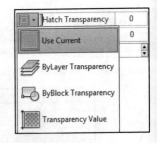

HATCH ANGLE

Pattern: The default angle of a pattern is **"0"**. If you change this angle the pattern will rotate relative to its original design.

User defined: Specify the actual angle of the hatch lines.

HATCH SCALE or SPACING

If Hatch Type **Pattern** is selected this value determines the scale of the Hatch Pattern. A value greater than **"1"** will increase the scale. A value less than **"1"** will decrease the scale.

If Hatch Type **User Defined** is selected this value determines the spacing between the hatch lines.

HATCH PROPERTIES....continued

ORIGIN Panel

You may specify where the Hatch will originate. Lower left, lower right, upper left, upper right, center or even the at the UCS Origin.

The Origin locations shown on the right are displayed when Hatch type, **Pattern**, **Solid** or **User Defined** are selected.

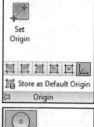

The Origin location **Centered** is displayed only when Hatch type **Gradient** is selected.

OPTIONS Panel

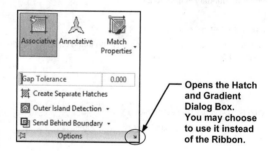

Opens the Hatch and Gradient Dialog Box. You may choose to use it instead of the Ribbon.

 Associative: If the Associative option is selected, the hatch set is associated to the boundary. This means if the boundary size is changed the hatch will automatically change to fit the new boundary shape.

 Annotative: AutoCAD will automatically adjusts the scale to match the current Annotative scale.

Continued on the next page...

HATCH PROPERTIES....continued

OPTIONS Panel....continued

MATCH PROPERTIES

Match Properties allows you to set the properties of the new hatch set by selecting an existing Hatch set. You may choose to "use the current origin" or "use the source hatch origin".

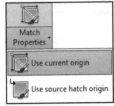

Use current origin: sets all properties <u>except the hatch origin</u>.

Use source hatch origin: sets all properties <u>including the hatch origin</u>.

GAP TOLERANCE

If the area you selected to hatch is not completely closed (gaps) AutoCAD will bridge the gap depending on the Gap tolerance. The Gap tolerance can be set to a value from 0 to 5000. Any gaps equal to or smaller than the value you specify are ignored and the boundary is treated as closed.

CREATE SEPARATE HATCHES

Controls whether HATCH creates a single hatch object or separate hatch objects when selecting several closed boundaries.

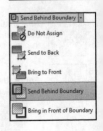

OUTER ISLAND DETECTION

These selections determine how Hatch recognizes internal objects.

SEND BEHIND BOUNDARY

These selections determine the draw order of the Hatch set.

HATCH TYPES

PATTERNS

PATTERNS

AutoCAD includes many previously designed Hatch Patterns. **Note:** Using Hatch Patterns will greatly increase the size of the drawing file. So use them conservatively. You may also purchase patterns from other software companies.

1. Select the Hatch Type **PATTERN**.

2. Select a Pattern from the list of hatch patterns displayed in the Pattern panel.

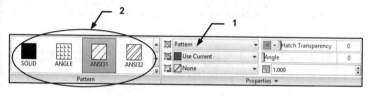

3. Select the Hatch Color, Boundary Background Color, Hatch Transparency.

4. **Angle:** A previously designed pattern has a default angle of **"0"**. If you change this Angle the pattern will rotate the pattern relative to its original design.
   It is important that you understand how to control the angle.

   For example: If **ANSI31** hatch pattern is used and the angle is set to **"45"** degrees, the pattern will rotate relative to its original design and the pattern will appear to be **"90"** degrees.

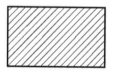

Original Pattern Design
Angle = 0

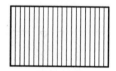

Pattern Design Rotated
Angle = 45

5. **Scale:** A value greater than **"1"** will increase the scale. A value less than **"1"** will decrease the scale. If the Hatch set is Annotative the scale will automatically adjust to the Annotative scale. But you might have to tweak the scale additionally to make it display exactly as you desire.

HATCH TYPES....continued

USER DEFINED

USER DEFINED
This Hatch Type allows you to simply draw continuous lines. No special pattern.
You specify the Angle of the lines and the Spacing between the lines.
Note: This Hatch type does not significantly increase the size of the drawing file.

1. Select the Hatch Type **USER DEFINED**.

2. Note the Pattern swatches are not displayed in the Pattern panel. (The Zigzag is only there because alphabetically it was in the same row as User Defined)

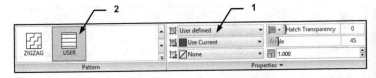

3. Select the Hatch Color, Boundary Background Color, Hatch Transparency.

4. **Angle:** Specify the actual angle that you desire from **"0"** to **"180"**.

   For example: If you want the lines to be on an angle of **"45"** degrees you would enter **"45"**. If you want the lines to be on an angle of **"90"** degrees you would enter **"90"**. (This is different from the angle for Patterns)

Angle = 45

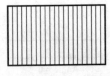

Angle = 90

5. **Spacing:** Specify the actual distance between each of the hatch lines.

HATCH TYPES....continued

SOLID

SOLID
If you would like to fill an area with a solid fill you should use Hatch type **Solid**.

1. Select **Solid** by selecting the Hatch Type **Solid** on the Properties Panel.
   or
   Select the **Solid** swatch on the Pattern Panel.

2. Select the Hatch **Color**.

Note: The Boundary Background Color is not available when using Hatch Type **Solid**.

3. Select **Transparency**.

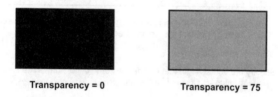

Transparency = 0　　　　**Transparency = 75**

4. **Angle:** Not available when using Hatch Type Solid.

5. **Scale:** Not available when using Hatch Type Solid.

HATCH TYPES....continued

GRADIENT

GRADIENT
Gradients are fills that gradually change from dark to light or from one color to another.
Gradient fills can be used to enhance presentation drawings, giving the appearance of
light reflecting on an object, or creating interesting backgrounds for illustrations.

Gradients are definitely fun to experiment with but you will have to practice to achieve
complete control. They will also greatly increase the size of the drawing file.

1. Select the Hatch Type **GRADIENT**.

2. Select a Gradient Pattern from the **9 GR_ patterns** displayed in the Pattern panel.

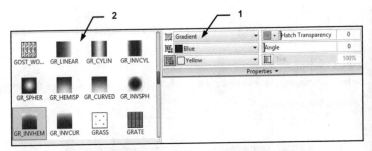

3. Select the Hatch **Color**. The Gradient can be <u>one color</u> or <u>two color</u>. If one color you
   can select the Tint and Shade of that color. (See step 6 below)

4. Select the Hatch **Transparency**.

5. **Angle:** Specify the actual angle that you desire from **"0"** to **"180"**. The pattern will
   rotate the pattern relative to its original design.

| Angle = 0 | Angle = 45 |

6. **Tint and Shade:** Tint and Shade is used when you are using only one-color gradient
   fill. Specify the tint or shade of the color selected in step 3 above.

EDITING HATCH

EDITING THE HATCH SET PROPERTIES

Editing a Hatch set properties is easy.
1. Simply select the Hatch set to edit.
2. The Hatch Editor Ribbon tab will appear.
3. Make new selections. Any changes are applied immediately.

CHANGING THE BOUNDARY

If the Hatch set is **Associative** (see 4-33) you may change the shape of the boundary and the Hatch set will conform to the new shape.
If the Hatch set is **Non-Associative** the Hatch set will not change.

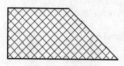
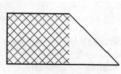

Original Boundary   **New Boundary**   **New Boundary**
                        with Associative Hatch   **with non-Associative Hatch**

TRIMMING HATCH

You may trim a hatch set just like any other object.

Before Trim   **After Trim**

MIRROR HATCH

An existing Hatch set can be mirrored. The boundary shape will automatically mirror. But you may control whether the Hatch pattern is mirrored or not.

To control the Hatch pattern mirror: (**Note:** Set prior to using the Mirror command)
1. Type: **mirrhatch <enter>**
2. Enter **0** or **1 <enter>**   0 = Hatch not Mirrored   1 = Hatch Mirrored

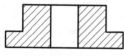
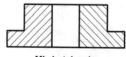

Mirrhatch = 0   **Mirrhatch = 1**

DRAWING LINES

A **Line** can be **one segment** or a **series of connected segments**.
But each segment is an individual object.

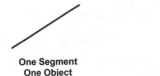

One Segment
One Object

Series of connected Segments
Five Objects

Start the **Line** command using one of the following methods:

Ribbon = Home Tab / Draw Panel /
or
Keyboard = L <enter>

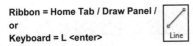

<u>**Lines** are drawn by specifying the locations for each endpoint.</u>

Move the cursor to the location of the **"first"** endpoint (**1**) then press the left mouse button and release. Move the cursor again to the **"next"** endpoint (**2**) and press the left mouse button. Continue locating **"next"** endpoints until you want to stop drawing lines.

<u>There are 2 ways to **Stop drawing a line**</u>:

Press the **<enter>** key or press the **<Space Bar>**

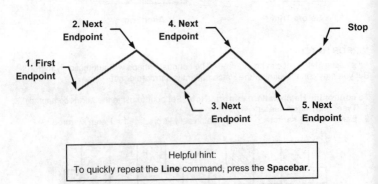

> Helpful hint:
> To quickly repeat the **Line** command, press the **Spacebar**.

Continued on the next page...

DRAWING LINES....continued

Horizontal and Vertical Lines

To draw a Line perfectly Horizontal or Vertical select the **Ortho** mode by selecting the **Ortho** button on the Status Bar or pressing the **F8** key.

 or

Try the following example:

1. Select the **Line** command. (Refer to the previous page)

2. Place the First endpoint anywhere in the drawing area.

3. **Turn Ortho ON** by selecting the **Ortho** button or **F8**. (The "Ortho" button will change to blue when ON.)

4. Move the cursor to the right and press the left mouse button to place the **next endpoint**. (The line should appear perfectly horizontal.)

5. Move the cursor down and press the left mouse button to place the **next endpoint**. (The line should appear perfectly vertical)

6. Now turn Ortho **OFF** by selecting the Ortho button. (The "Ortho" button will change to gray when OFF.)

7. Now move the cursor up and to the right on an angle (the line should move freely now) and press the left mouse button to place the **next endpoint**.

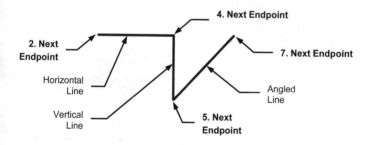

Ortho can be turned ON or OFF at any time while you are drawing. It can also be turned ON or OFF temporarily by holding down the **Shift** key. Release the Shift key to resume.

Continued on the next page...

DRAWING LINES....continued

Closing Lines

If you have drawn 2 or more line segments, the <u>endpoint of the last line segment</u> can be connected automatically to the <u>first endpoint</u> using the **Close** option.

Try the following example:

1. Select the **Line** command.

2. Place the **First endpoint**.

3. Place the **next endpoint**.

4. Place the **next endpoint**.

5. Type **C <enter>**

   Or

5. <u>Press the right mouse button</u> and select **Close** from the **Shortcut** menu.

3. Next Endpoint

2. First Endpoint

4. Next Endpoint

5. Automatically Closes

What is the Shortcut Menu?

The <u>Shortcut menu</u> gives you quick access to command options.

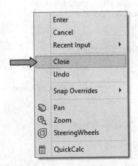

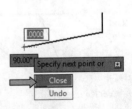

Using the Shortcut menu:

Press the right mouse button.
The shortcut menu will appear.
Select an option.

Using the Dynamic Input down arrow:

You may use the right mouse button or press the down arrow ↓ and the options will appear below the Dynamic Input prompt.

POINT

Points are used to locate a point of reference or location. A **Point** may be represented by one of many **Point Styles** shown below in the **Point Style dialog box**.

The only object snap option that can be used with Point is **Node**.

HOW TO USE THE POINT COMMAND

1. Select the **POINT** command using one of the following:

   Ribbon = Home Tab / Draw Panel /
   or
   Keyboard = PO <enter>

2. The following prompts will appear on the command line:

   Command: _point
   Current point modes: PDMODE=3 PDSIZE=0.000
   Specify a point: *place the point location*
   Specify a point: *place another point or press the "ESC" key to stop*

HOW TO <u>SELECT</u> A "POINT STYLE"

1. Open the Point Style dialog box:

   Ribbon = Home Tab / Utilities Panel ▼ / Point Style
   or
   Keyboard = ddtype <enter>

2. The Point Style dialog box will appear.

3. Select a point style tile.

4. Select the **OK** button.

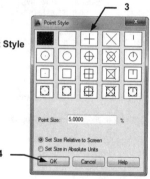

Point Size:

Set Size Relative to Screen
Sets the point display size as a percentage of the screen size. The point display does not change when you zoom in or out.

Set Size in Absolute Units
Sets the point display size as the actual units you specify under Point Size. Points are displayed larger or smaller when you zoom in or out.

POLYGON

A polygon is an object with multiple edges (flat sides) of equal length. You may specify from 3 to 1024 sides. A polygon appears to be multiple lines but in fact it is one object. You can specify the <u>center and a radius</u> or the <u>edge length</u>. The <u>radius</u> size can be specified <u>Inscribed</u> or <u>Circumscribed</u>.

CENTER, RADIUS METHOD

1. Select the **POLYGON** command using one of the following:

   Ribbon = Home Tab / Draw Panel /
   or
   Keyboard = POL <enter>

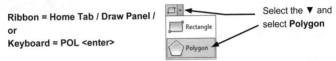

Select the ▼ and select **Polygon**

2. The following prompts will appear on the command line:

_polygon Enter number of sides <4>: *type number of sides <enter>*
Specify center of polygon or [Edge]: *specify the center location (P1)*
Enter an option [Inscribed in circle/Circumscribed about circle]<I>: *type I or C <enter>*
Specify radius of circle: *type radius or locate with cursor. (P2)*

Note:
The dashed circle is shown only as a reference to help you visualize the difference between Inscribed and Circumscribed. Notice that the radius is the same (2") but the Polygons are different sizes. Selecting Inscribed or Circumscribed is important.

INSCRIBED

CIRCUMSCRIBED

EDGE METHOD

1. Select the **POLYGON** command using one of the options shown above.

2. The following prompts will appear on the command line:

_polygon Enter number of sides <4>: *type number of sides <enter>*
Specify center of polygon or [Edge]: *type E <enter>*
Specify first endpoint of edge: *place first endpoint of edge (P1)*
Specify second endpoint of edge: *place second endpoint of edge (P2)*

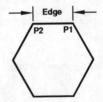

POLYLINES

A **POLYLINE** is very similar to a LINE. It is created in the same way a line is drawn.
It requires first and second endpoints. But a POLYLINE has additional features, as
follows:

1. A **POLYLINE** is ONE object, even though it may have many segments.
2. You may specify a specific width to each segment.
3. You may specify a different width to the start and end of a polyline segment.

Select the **POLYLINE** Command using one of the following:

Ribbon = Home Tab / Draw Panel /
or
Keyboard = PL <enter>

THE FOLLOWING ARE EXAMPLES OF POLYLINES WITH WIDTHS ASSIGNED.

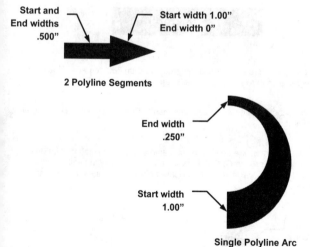

Start and
End widths
.500"

Start width 1.00"
End width 0"

2 Polyline Segments

End width
.250"

Start width
1.00"

Single Polyline Arc

Refer to the next page for more polyline options.

POLYLINES....continued

OPTIONS:

WIDTH
Specify the starting and ending width.

You can create a tapered polyline by specifying different starting and ending widths.

Starting width 1.00" — **Ending width 0"**

HALFWIDTH
The same as Width except the starting and ending halfwidth specifies half the width rather than the entire width.

ARC
This option allows you to create a circular polyline less than 360 degrees. You may use (2) 180 degree arcs to form a full circular shape.

CLOSE
The close option is the same as in the Line command. Close attaches the last segment to the first segment.

LENGTH
This option allows you to draw a polyline at the same angle as the last polyline drawn. This option is very similar to the **OFFSET** command. You specify the first endpoint and the length. The new polyline will automatically be drawn at the same angle as the previous polyline.

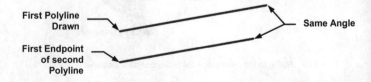

First Polyline Drawn — **Same Angle**

First Endpoint of second Polyline

POLYLINES....continued

CONTROLLING THE FILL MODE

If you turn the **FILL** mode **OFF** the polylines will appear as shown below.

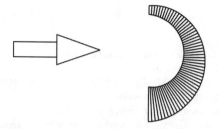

How to turn FILL MODE on or off.

1. Command: *type FILL <enter>*
2. Enter mode [On / Off] <ON>: *type ON or Off <enter>*
3. Command: *type REGEN <enter> or select: View / Regen*

EXPLODING A POLYLINE

NOTE: If you **Explode** a **POLYLINE** it loses its width and turns into a regular line as shown below.

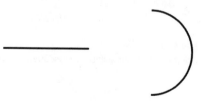

POLYLINES....continued

The following is an example of how to draw a polyline with Width.

1. Select the **POLYLINE** Command using one of the following:

    Ribbon = Home Tab / Draw Panel /
    or
    Keyboard = PL <enter>

    Polyline

    Command: _pline
2. Specify start point: *place the first endpoint of the line*
    Current line-width is 0.000
3. Specify next point or [Arc/Halfwidth/Length/Undo/Width]: *select width option*
4. Specify starting width <0.000>: *1 <enter>*
5. Specify ending width <0.000>: *1 <enter>*
6. Specify next point or [Arc/Close/Halfwidth/Length/Undo/Width]: *select Length*
    option

7. Specify Length: *3 <enter>*
8. Specify next point or [Arc/Halfwidth/Length/Undo/Width]: *select width option*
9. Specify starting width <0.000>: *2 <enter>*
10. Specify ending width <0.000>: *.5 <enter>*
11. Specify next point or [Arc/Close/Halfwidth/Length/Undo/Width]: *select Length*
    option

12. Specify Length: *2.75 <enter>*
13. Specify next point or [Arc/Halfwidth/Length/Undo/Width]: *select width option*
14. Specify starting width <0.000>: *.5 <enter>*
15. Specify ending width <0.000>: *1 <enter>*
16. Specify next point or [Arc/Close/Halfwidth/Length/Undo/Width]: *select Length*
    option

17. Specify Length: *2.50 <enter>*
18. Specify next point or [Arc/Close/Halfwidth/Length/Undo/Width]: *press <enter> to*
    stop

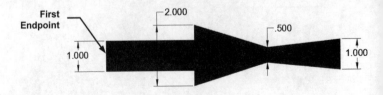

EDITING POLYLINES

The **POLYEDIT** command allows you to make changes to a polyline's option, such as the width. You can also change a regular line into a polyline and JOIN the segments.

Note: If you select a line that is **NOT a POLYLINE**, the prompt will ask if you would like to turn it into a POLYLINE.

1. Select the **POLYEDIT** command using one of the following:

Ribbon = Home Tab / Modify Panel /
or
Keyboard = PE <enter>

Note: You may modify "Multiple" polylines simultaneously.

2. PEDIT Select polyline or [Multiple]: *select the polyline to be edited or "M"*

3. Enter an option [Close/Join/Width/Edit vertex/Fit/Spline/Decurve/Ltypegen/Undo/Reverse]: *select an Option (descriptions of each are listed below.)*

OPTIONS

CLOSE

CLOSE connects the last segment with the first segment of an Open polyline.
AutoCAD considers a polyline open unless you use the **"Close"** option to connect the segments originally.

Open

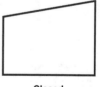

Closed

OPEN

OPEN removes the closing segment, but only if the **CLOSE** option was used to close the polyline originally.

EDITING POLYLINES....continued

OPTIONS:

JOIN

The JOIN option allows you to join individual polyline segments into one polyline.
The segments must have matching endpoints.

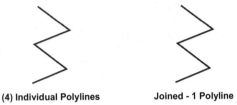

(4) Individual Polylines　　　　**Joined - 1 Polyline**

WIDTH

The WIDTH option allows you to change the width of the polyline.
But the entire polyline will have the same width.

EDIT VERTEX

This option allows you to change the starting and ending width of each segment
individually.

SPLINE

This option allows you to change
straight polylines to curves.

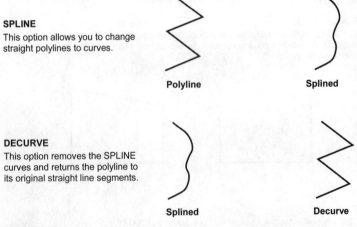

Polyline　　　　**Splined**

DECURVE

This option removes the SPLINE
curves and returns the polyline to
its original straight line segments.

Splined　　　　**Decurve**

REVERSE

This option reverses the direction. The start point becomes the end point and vice
versa.

RECTANGLE

A Rectangle is a closed rectangular shape. It is one object not 4 lines.
You can specify the length, width, area, and rotation options.
You can also control the type of corners on the rectangle—fillet, chamfer, or square and
the width of the Line.

First, let's start with a simple Rectangle using the cursor to select the corners.

1. Start the **RECTANGLE** command by using one of the following:

   Ribbon = Home Tab / Draw Panel /
   or
   Keyboard = REC <enter>

2. The following will appear on the command line:

   Command: _rectang
   Specify first corner point or [Chamfer/Elevation/Fillet/Thickness/Width]:

3. Specify the location of the first corner by moving the cursor to a location (**P1**) and
   then press the left mouse button.

   The following will appear on the command line:

   Specify other corner point or [Area / Dimensions / Rotation]:

4. Specify the location of the **diagonal** corner (**P2**) by moving the cursor diagonally
   away from the first corner (**P1**) and pressing the left mouse button.

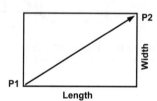

OR

4. Type **D <enter>** (or click on the blue letter **"D"**)

   Specify length for rectangles <0.000>: *Type the desired length <enter>.*

   Specify width for rectangles <0.000>: *Type the desired width <enter>.*

   Specify other corner point or [Dimension]: *move the cursor up, down, right or left
   to specify where you want the second corner relative to the first corner and
   then press <enter> or press left mouse button.*

RECTANGLE....continued

OPTIONS: Chamfer, Fillet and Width
Note: the following options are **only** available **before** you place the **first corner** of the Rectangle.

CHAMFER
A chamfer is an angled corner. The Chamfer option automatically draws all 4 corners with chamfers simultaneously and all the same size. You must specify the distance for each side of the corner as **distance 1** and **distance 2**.

Example: A Rectangle with dist1 = .50 and dist2 = .25

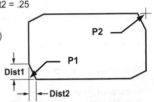

1. Select the **RECTANGLE** command
2. Type **C <enter>** (or click on the blue letter **"C"**)
3. Enter **.50** for the first distance
4. Enter **.25** for the second distance
5. Place the first corner (**P1**)
6. Place the diagonal corner (**P2**)

FILLET
A fillet is a rounded corner. The fillet option automatically draws all 4 corners with fillets (all the same size). You must specify the radius for the rounded corners.

Example: A Rectangle with .50 radius corners.

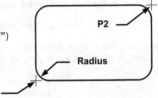

1. Select the **RECTANGLE** command
2. Type **F <enter>** (or click on the blue letter **"F"**)
3. Enter **.50** for the radius.
4. Place the first corner (**P1**)
5. Place the diagonal corner (**P2**)

Note: You must set Chamfer and Fillet back to "0" before defining the width. Unless you want fat lines and Chamfered or Filleted corners.

WIDTH
Defines the width of the rectangle lines.
Note: Do not confuse this with the "Dimensions" Length and Width.
Width makes the lines appear fatter.

Example: A Rectangle with a width of .50

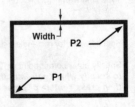

1. Select the **RECTANGLE** command
2. Type **W <enter>** (or click on the blue letter **"W"**)
3. Enter **.50** for the width.
4. Place the first corner (**P1**)
5. Place the diagonal corner (**P2**)

RECTANGLE....continued

OPTIONS: Area and Rotation
Note: the following options are available **After** you place the **first corner** of the Rectangle.

AREA
Creates a Rectangle using the AREA and either a LENGTH or a WIDTH. If the Chamfer or Fillet option is active, the area includes the effect of the chamfers or fillets on the corners of the rectangle.

Example: A Rectangle with an Area of 6 and a Length of 2.

1. Select the **RECTANGLE** command
2. Place the first corner (**P1**)
3. Type **A <enter>** for Area. (or click on blue "**A**")
4. Enter **6 <enter>** for the Area
5. Select **L <enter>** for length option (or click on blue "**L**")
6. Enter **2 <enter>** for the length
   (The width will automatically be calculated)

ROTATION
You may select the desired rotation angle **After** you place the **first corner** and **Before** you place the **second corner**. The base point (pivot point) is the first corner.

Note: All new rectangles within the drawing will also be rotated unless you reset the rotation to **0**. This option will not effect rectangles already in the drawing.

Example: A Rectangle with a rotation angle of 45 degrees.

1. Select the **RECTANGLE** command
2. Place the first corner (**P1**)
3. Type **R <enter>** for rotation. (or click on blue "**R**")
4. Enter **45 <enter>**
5. Place the diagonal corner (**P2**)

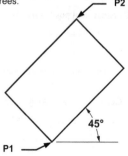

CREATING A REVISION CLOUD

When you make a revision to a drawing it is sometimes helpful to highlight the revision for someone viewing the drawing. A common method to highlight the area is to draw a "Revision Cloud" around the revised area. This can be accomplished easily with the "Revision Cloud" command.

The Revision Cloud command creates a series of sequential arcs to form a cloud shaped object. You set the minimum and maximum arc lengths. (Maximum arc length cannot exceed three times the minimum arc length. Example: Min = 1, Max can be 3 or less) If you set the minimum and maximum different lengths the arcs will vary in size and will display an irregular appearance.

Min & Max same length   **Min & Max different length**

To draw a Revision Cloud you specify the start point with a left click then drag the cursor to form the outline. AutoCAD automatically draws the arcs. When the cursor gets very close to the start point, AutoCAD snaps the last arc to the first arc and closes the shape.

1. Select the **REVISION CLOUD** command using one of the following:

   Ribbon = Home Tab / Draw Panel ▼ /
   or
   Keyboard = revcloud <enter>

   Command: _revcloud
   Minimum arc length: .50   Maximum arc length: .50   Style: Normal

2. Specify start point or [Arc length/Object/Style] <Object>: *Select "Arc length"*

3. Specify minimum length of arc <.50>: *Specify the minimum arc length*

4. Specify maximum length of arc <.50>: *Specify the maximum arc length*

5. Specify start point or [Arc length/Object/Style] <Object>: *Place cursor at start location & left click.*

6. Guide crosshairs along cloud path...*Move the cursor to create the cloud outline.*

7. Revision cloud finished. *When the cursor approaches the start point, the cloud closes automatically.*

CONVERT A CLOSED OBJECT TO A REVISION CLOUD

You can convert a closed object, such as a circle, ellipse, rectangle or closed polyline to a revision cloud. The original object is deleted when it is converted.

(If you want the original object to remain, in addition to the new rev cloud, set the variable **"delobj"** to **"0"**. The default setting is **"1"**.)

1. Draw a closed object such as a circle.

2. Select the **REVISION CLOUD** command using one of the following:

Ribbon = Home Tab / Draw Panel ▼ /
or
Keyboard = revcloud <enter>

Command: _revcloud
Minimum arc length: .50   Maximum arc length: .50   Style: Normal

3. Specify start point or [Arc length/Object/Style] <Object>: **Select "Arc length"**

4. Specify minimum length of arc <.50>: **Specify the minimum arc length**

5. Specify maximum length of arc <.50>: **Specify the maximum arc length**

6. Specify start point or [Arc length/Object/Style] <Object>: **Select "Object"**

7. Select object: **Select the object to convert**

8. Select object: Reverse direction [Yes/No] <No>: **Select Yes or No**
   Revision cloud finished.

Reverse direction?

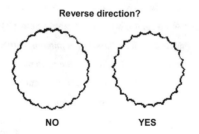

NO   YES

NOTE:
The Match Properties command will not match the arc length from the source cloud to the destination cloud.

REVISION CLOUD STYLE

You may select one of 2 styles for the Revision Cloud; **Normal** or **Calligraphy**.
Normal will draw the cloud with one line width.
Calligraphy will draw the cloud with variable line widths to appear as though you used a chiseled calligraphy pen.

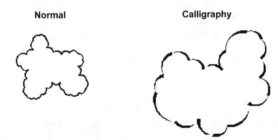

Normal   Calligraphy

1. Select the **REVISION CLOUD** command using one of the following:

   Ribbon = Home Tab / Draw Panel ▼ /
   or
   Keyboard = revcloud <enter>

   Command: _revcloud
   Minimum arc length: .50   Maximum arc length: 1.00   Style: Normal

2. Specify start point or [Arc length/Object/Style] <Object>: *Select "Style"<enter>*

3. Select arc style [Normal/Calligraphy] <Calligraphy>: *Select "N or C"<enter>*

4. Specify start point or [Arc length/Object/Style] <Object>: *Select "Arc length"*

5. Specify minimum length of arc <.50>: *Specify the minimum arc length*

6. Specify maximum length of arc <1.00>: *Specify the maximum arc length*

7. Specify start point or [Object] <Object>: *Place cursor at start location & left click*

8. Guide crosshairs along cloud path...*Move the cursor to create the cloud outline*

9. Revision cloud finished. *When the cursor approaches the start point, the cloud closes automatically*

Section 5
How to.....

ADD A PRINTER / PLOTTER

The following are step-by-step instructions on how to configure AutoCAD for your printer or plotter. These instructions assume you are a single system user. If you are networked or need more detailed information, please refer to your AutoCAD Help Index.

Note: You can configure AutoCAD for multiple printers. Configuring a printer makes it possible for AutoCAD to display the printing parameters for that printer.

A. Type: *plottermanager <enter>*

or **Application menu / Print / Manage Plotters** ⟶

B. Select **"Add-a-Plotter Wizard"**

Add-A-Plotter Wizard
Shortcut
1.20 KB

C. Select the **"Next"** button.

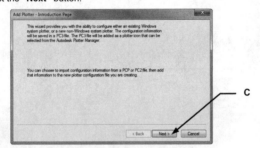

D. Select **"My Computer"** then **Next**.

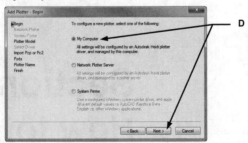

Continued on the next page...

E. Select the **Manufacturer** and the specific **Model** desired then **Next**.

(If you have a disk with the specific driver information, put the disk in the disk drive and select **"Have disk"** button, then follow instructions.)

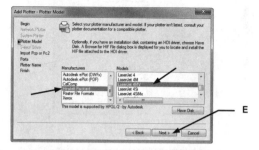

F. Select the **"Next"** button.

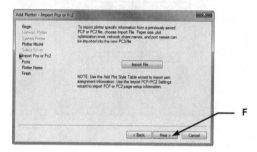

G1. Select **"Plot to a port"**.

G2. Then select **"Next"**.

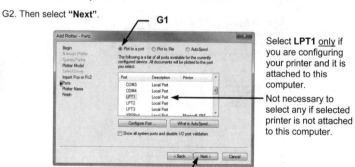

Select **LPT1** <u>only</u> if you are configuring your printer and it is attached to this computer.

Not necessary to select any if selected printer is not attached to this computer.

Continued on the next page...

H. The Printer name that you previously selected should appear. Then select **"Next"**.

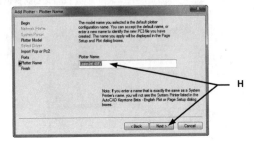

I. Select the **"Edit Plotter Configuration..."** box.

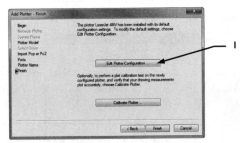

J. Select:
1. Device and Document Settings tab.
2. Media: Source and Size
3. Size: (Select the appropriate size for your printer / plotter)
4. **OK** box.

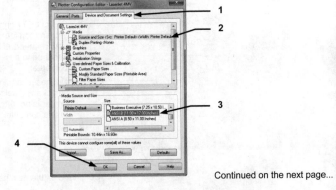

Continued on the next page...

K. Select **"Finish"**.

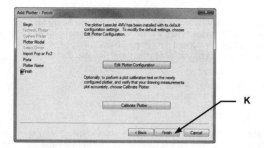

— K

L. Type: *Plottermanager <enter>* again.

Is the printer / plotter there?

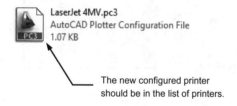

The new configured printer
should be in the list of printers.

HOW TO CREATE A PAGE SETUP

When you select a layout tab for the first time the **Page Setup Manager** will appear. The Page Setup Manager allows you to select the **printer/plotter** and **paper size**. These specifications are called the **"Page Setup"**. This page setup will be saved to that layout tab so it will be available when ever you use that layout tab.

1. **Open** the drawing you wish to plot.
   (The drawing must be displayed on the screen.)

2. Select a **Layout tab**.

Note: If the "Page Setup Manager" dialog box shown below does not appear automatically, right click on the Layout tab and select Page Setup Manager.

3. Select the **New...** button.

Check this box if the Page Setup Manager did not appear automatically when you selected a Layout tab.

Yours will be different. It may even display **"None"** That's OK for now.

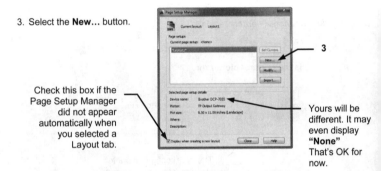

4. Select **<Default output device>** in the **Start with:** list.

5. Enter the New page setup name: **Setup A**

6. Select **OK** button.

(I am assuming that your computer is attached to a printer. If not select Layout1)

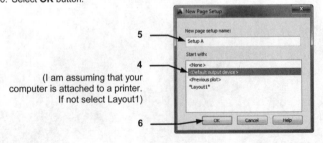

Continued on the next page...

HOW TO CREATE A PAGE SETUP....continued

This is where you will select the **printer / plotter**, **paper size** and the **plot offset**.

7. Select the **Printer / Plotter**
   Note: Your current system printer should already be displayed here. If you prefer another select the down arrow and select from the list. If the preferred printer is not in the list you must configure the printer. (Refer to Add a Printer in Section 5.)

8. Select the **Paper Size**

9. Select **Plot Offset**

Notice the name you entered is now displayed as the page setup name.

7
(Yours will be different)

8
(Yours may state 8-1/2 x 11)

9
(Should remain at 0.00000 0.00000)

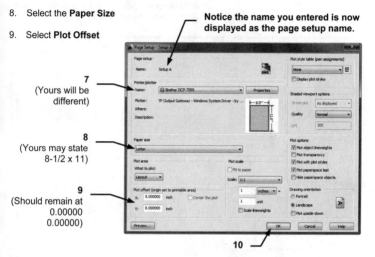

10

10. Select the **OK** button.

11. Select the **Page Setup** (Setup A).

12. Select the **Set Current** button.

13. Select the **Close** button.

12

11

13

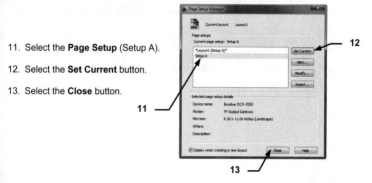

Continued on the next page...

You should now have a sheet of paper displayed on the screen.
This sheet is the size you specified in the "Page Setup".
This sheet is in front of Model Space.

The dashed line represents the maximum printing area for the printing device that you selected. Any object outside of this area will not print.

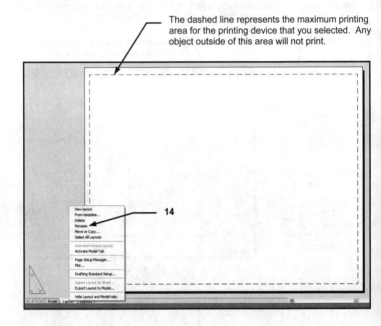

Rename the Layout tab

14. Right click on the active Layout tab and select **Rename** from the list.

15. Enter the new Layout name: **A Size <enter>**

VIEWPORTS

Viewports are only used in Paper Space (Layout tab).
Viewports are holes cut into the sheet of paper displayed on the screen in Paper Space.
Viewports frames are objects. They can be moved, stretched, scaled, copied and erased. You can have multiple Viewports, and their size and shape can vary.

Note: It is considered good drawing management to create a layer for the Viewport "frames" to reside on. This will allow you to control them separately; such as setting the viewport layer to "No plot" so it will not be plotted out.

HOW TO CREATE A VIEWPORT

1. First, create a drawing in **Model Space** (Model tab) and save it.

2. Select the **"Layout1"** tab.

   When the **"Page Setup Manager"** dialog box appears, select the **New** button. Then you will select the <u>Printing device</u> and <u>paper size</u> to plot on.
   (Refer to "How to Create a Page setup")

3. You are now in Paper Space. Model Space appears to have disappeared, because a blank paper is now in front of Model Space, preventing you from seeing your drawing. You designated the size of this sheet in the "page setup" mentioned in #2 above. (The Border, title block and notes will be drawn on this paper.)

You designated the size of this sheet in the Page Setup

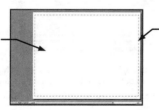

The dashed line represents the printing limits for the printer that you selected in the Page Setup

Continued on the next page...

5-9

VIEWPORTS....continued

4. Draw a border, title block and notes in **Paper Space** (Layout).

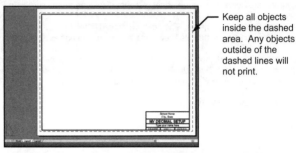

Keep all objects inside the dashed area. Any objects outside of the dashed lines will not print.

Now you will want to see the drawing that is in Model Space.

5. Select layer **"Viewport"** (You want the viewport frame to be on layer viewport)

6. Select the **VIEWPORT** command using one of the following:

Ribbon = Layout Tab / Layout Viewports Panel /
or
Keyboard = MV <enter>

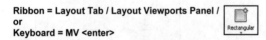

7. Draw a rectangular shaped Viewport "frame" by placing the location for the "first corner" and then the "opposite corner" using the cursor. (Similar to drawing a Rectangle, but **do not** use the Rectangle command. You must remain in the **MV** command)

Opposite corner of Viewport

First corner of Viewport

You should now be able to look through the Paper Space sheet to Model Space and see your drawing because you just cut a rectangular shaped hole in the sheet.

Note: Now you may go back to Model Space or return to Paper Space, simply by selecting the tabs, model or layout.

(Make sure your grids are **ON** in Model Space and **OFF** in Paper Space. Otherwise you will have double grids)

5-10

HOW TO REACH INTO A VIEWPORT

Here are the rules:

1. You have to be in Paper Space (layout tab) and at least one viewport must have been created.

2. You have to be <u>inside a Viewport</u> to manipulate the scale or position of the drawing that you see in that Viewport.

How to reach into a viewport to manipulate the display.

<u>First, select a layout tab and cut a viewport</u>

At the bottom right of the screen on the status bar there is a button that either says Model or Paper. This button displays which space you are in currently. `PAPER`

When the button is PAPER you are working on the Paper sheet that is in front of Model Space. You may cut a viewport, draw a border, title block and place notes.

<u>If you want to reach into a viewport to manipulate the display</u>, double click inside of the viewport frame. Only one viewport can be activated at one time. The active viewport is indicated by a heavier viewport frame. (Refer to the illustration on the previous page. The viewport displaying the doors is active). `MODEL`
Also, the Paper button changed to Model.

While you are inside a viewport you may manipulate the scale and position of the drawing displayed. To return to the Paper surface click on the word Model and it will change to Paper. `PAPER`   You may now work on the paper surface.

Note: Do not confuse the <u>Model / Layout tabs</u> with the <u>MODEL / PAPER button</u>.

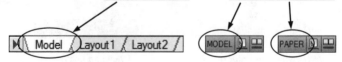

Here is the difference.

The **Model / Layout tabs** shuffle you from the actual drawing area (model space) to the Layout area (paper space). It is sort of like if you had 2 stacked pieces of paper and when you select the Model tab the drawing would come to the front and you could not see the layout. When you select the Layout tab a blank sheet would come to the front and you would not see model space......unless you have a viewport cut.

The **MODEL / PAPER button** allows you to work in model space or paper space without leaving the layout tab. No flipping of sheets. You are either on the paper surface or in the viewport reaching through to model space.

HOW TO LOCK A VIEWPORT

After you have manipulated the drawing within each viewport, to suit your display needs, you will want to **LOCK** the viewport so the display can't be changed accidentally. Then you may zoom in and out and you will not disturb the display.

1. Make sure you are in **Paper Space**.

2. Click once on a **Viewport Frame**.

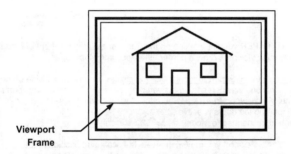

Viewport Frame

3. Click on the **Open Lock tool** located in the lower right corner of drawing area.

   The icon will change to a **Closed Lock tool**.

Viewport Unlocked   **Viewport Locked**

Now, any time you want to know if a Viewport is locked or unlocked just glance down to the <u>*Lock tool*</u> *shown above.*

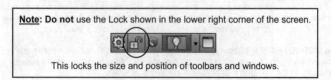

> **Note:** **Do not** use the Lock shown in the lower right corner of the screen.
>
> This locks the size and position of toolbars and windows.

USING THE LAYOUT

Now that you have the correct paper size on the screen, you need to do a little bit more to make it useful.

The next step is to:

1. Add a Border, Title Block and notes in Paper Space.
2. Cut a viewport to see through to Model Space.

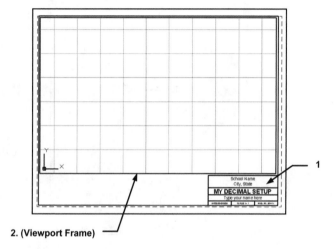

1

2. (Viewport Frame)

3. Adjust the scale of the viewport as follows.
   A. You must be in **Paper Space**.
   B. Click on the **Viewport Frame**.
   C. **Unlock** Viewport if it is locked.
   D. Select the **Viewport Scale down arrow ▼**.
   E. Select **Scale** (List shown on next page)

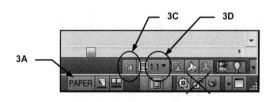

3C   3D

3A

Continued on the next page...

USING THE LAYOUT....continued

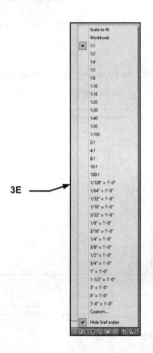

3E

4. **Lock** the Viewport.

Now you may zoom as much as you desire and it will not affect the adjusted scale.

CREATE A TEMPLATE

1. Set up the drawing and consider the following:

   - **Drawing Limits**
   - **Layers**
   - **Dimension Styles**
   - **Title Blocks**
   - **Drawing Units**
   - **Text Styles**
   - **Layouts**

2. Select and click on the **"Application Menu ▼ "**

 2013   2014

3. Select **Save As / " ▶ "** (Click on the arrow, not the words "Save As")

4. Select **"Drawing Template"**

   (**Note:** *There are slight differences in the button icons that are used in AutoCAD 2013 and AutoCAD 2014, but both have identical functions. Both versions are shown below*).

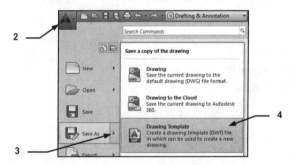

AutoCAD 2013 Drawing Template Icon

> **Drawing Template**
> Create a drawing template (DWT) file in which can be used to create a new drawing.

AutoCAD 2014 Drawing Template Icon

> **Drawing Template**
> Create a drawing template (DWT) file in which can be used to create a new drawing.

Continued on the next page...

CREATE A TEMPLATE....continued

5. Type the new file name in the **File Name** box.
   Do not type the extension **.dwt**, AutoCAD will add it automatically.

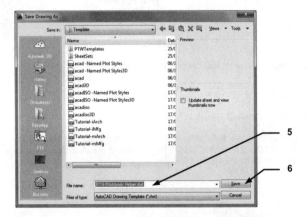

6. Select the **Save** button.

7. Type a **Description** for the Template.

8. Select the **OK** button.

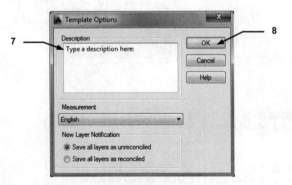

Now you have a Template to use for future drawings.

Using a Template as a master setup drawing is very good CAD management.

USING A TEMPLATE

TO USE A TEMPLATE

1. Select the **NEW** tool from the **Quick Access Toolbar**.

2. Select **Drawing Template [*.dwt]** from the "<u>Files of type</u>" if not already selected.

3. Select the **Template name** from the list of templates.

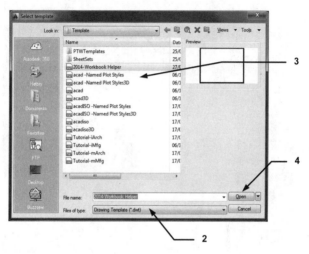

4. Select the **Open** button.

OPENING AN EXISTING DRAWING FILE

Opening an Existing Drawing File means that you would like to open, on to the screen, a drawing that has been previously created and saved. Usually you are opening it to continue working on it or you need to make some changes.

1. Select the **OPEN** tool on the **Quick Access Toolbar**.

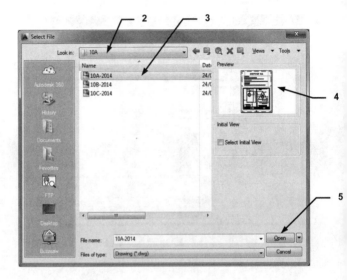

2. Locate the Directory and Folder in which the file had previously been saved.

3. Select the File that you wish to OPEN.

4. A **Thumbnail Preview Image** of the file selected is displayed in the Preview area.

5. Select the **Open** button.

OPEN MULTIPLE FILES 2014

(**Not** available in AutoCAD 2013)

The new **File Tabs** tool allows you to have multiple drawings open at the same time. If the File Tabs tool is switched on, you can open existing saved drawings or create new ones.

The **File Tabs** tool is located on the **User Interface** panel of the **View** tab, and is a Light Blue color when switched on.

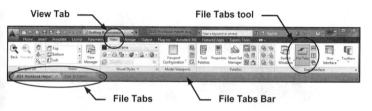

View Tab —   File Tabs tool —

File Tabs   File Tabs Bar

How to open an existing saved drawing from the Files Tab.

1. Right mouse click on the "**+**" icon.

2. Select **Open** from the menu.

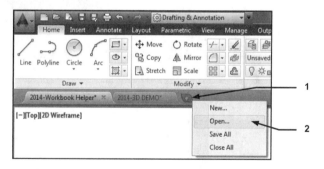

3. Locate the Directory and Folder for the previously saved file.

4. Select the File you wish to open.

5. Select the **Open** button.

Continued on the next page...

How to open a new drawing from the Files Tab.

1. Left mouse click on the "**+**" icon.

2. Select **Drawing Template (*.dwt)** from the **Files of type** drop-down list.

3. Select the Template you require.

4. Select the **Open** button.

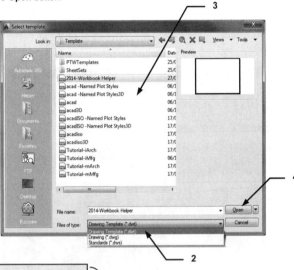

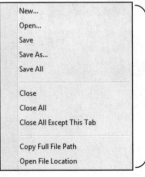

Note:
If you right mouse click on any **File Tab** a menu appears with various options, including closing all open drawing tabs except the one you just clicked on.

Continued on the next page...

OPEN MULTIPLE FILES....continued

<u>2014</u>

The new File Tabs drawing previews allow you to quickly change between open drawings. If you hover your mouse over any open File Tab, a preview of the **Model** and the **Layout** tabs are displayed. You can click on any of the previews to take you to that particular open drawing or view.

Hover the mouse over any
File Tab to see a preview of
the **Model** and **Layout** Tabs

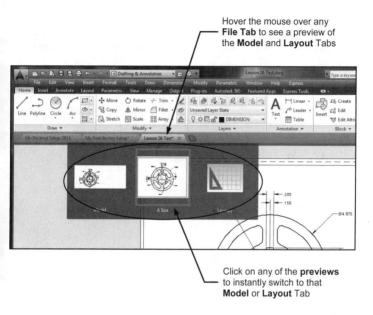

Click on any of the **previews**
to instantly switch to that
Model or **Layout** Tab

If an **asterisk** * is displayed on a File Tab it means that particular drawing has not been saved since it was last modified. The asterisk will disappear when the drawing has been saved.

**These two drawings are
showing the asterisk and
have not been saved**

SAVING A DRAWING FILE

After starting a new drawing, it is best practice to save it immediately. Learning how to save a drawing correctly is almost more important than making the drawing. If you can't save correctly, you will lose the drawing and hours of work.

There are two commands for saving a drawing: **Save** and **Save As**.
I prefer to use **Save As**.

The **Save As** command always pauses to allow you to choose where you want to store the file and what name to assign to the file. This may seem like a small thing, but it has saved me many times from saving a drawing on top of another drawing by mistake.

The **Save** command will automatically save the file either back to where you retrieved it or where you last saved a previous drawing. Neither may be the correct destination. And may replace a file with the same name. So play it safe, use **Save As** for now.

1. Select the **SAVEAS** command using one of the following:

> **Quick Access Toolbar =**
> or
> **Application Menu = Save As / Drawing**
> or
> **Keyboard = SA <enter> Save As**

2. **Select the "Save In" location.**
   (This is where the file will be saved.)

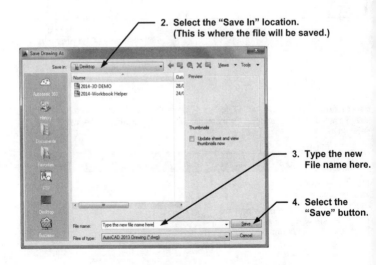

3. **Type the new File name here.**

4. **Select the "Save" button.**

AUTOMATIC SAVE

If you turn the automatic save option **ON**, your drawing is saved at specified time intervals. These temporary files are automatically deleted when a drawing closes normally. The default save time is every 10 minutes. You may change the save time Intervals and where you would prefer the Automatic Save files to be saved.

How to set the Automatic Save intervals

1. Type: *options <enter>*

2. Select the **Open and Save** tab.

3. Enter the desired **minutes between saves**.

4. Select the **OK** button.

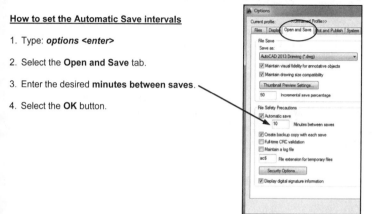

How to change the Automatic Save location

1. Type: *options <enter>*

2. Select the **Files** tab.

3. Locate the **Automatic Save File Location** and click on the **+** to display the **path**.

4. Double click on the path.

5. Browse to locate the Automatic Save Location desired and highlight it.

6. Select **OK**.

(The browse box will disappear and the new location path should be displayed under the Automatic Save File Location heading)

7. Select **OK** to accept the change.

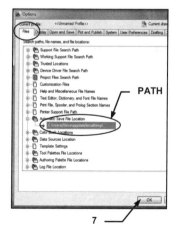

PATH

7

BACK UP AND RECOVER

When you save a drawing file, Autocad creates a file with a **".dwg"** extension.
For example, if you save a drawing as **12b**, Autocad saves it as **12b.dwg**.
The next time you save that same drawing, Autocad replaces the old with the new and
renames the old version **12b.bak**. The old version is now a back up file.
(Only 1 backup file for each drawing file is stored.)

How to open a back up file:
You can't open a **".bak"** file.
It must first be renamed with a **".dwg"** file extension.

How to view the list of back up files:
The backup files will be saved in the same location as the drawing file.
You must use Windows Explorer to locate the **".bak"** files.

How to rename a back up file:
1. Right click on the file name.
2. Select **"Rename"**.
3. Change the **".bak"** extension to **".dwg"** and press **<enter>**.

RECOVERING A DRAWING

In the event of a program failure or a power failure any open files should be saved
automatically.

When you attempt to re-open the drawing the **Drawing Recovery Manager** will
display a list of all drawing files that were open at the time of a program or system
failure. You can preview and open each **".dwg"** or **".bak"** file to choose which one
should be saved as the primary file.

STARTING A NEW DRAWING

Starting A New Drawing means that you want to start with a previously created Template file

Note: *Do not use the **New** tool if you want to **open** an **existing drawing**.*

<u>**HOW TO START A NEW DRAWING**</u>

1. Select the **NEW** tool from the **Quick Access Toolbar**.

2. Select the **Template file** from the list of templates.

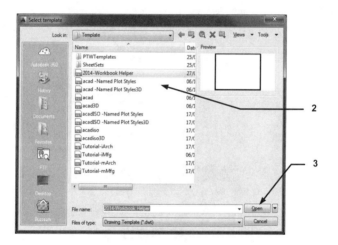

3. Select the **Open** button.

HOW TO CREATE AN AUTODESK ACCOUNT

In AutoCAD you can connect directly to the Autodesk 360 Cloud for online file sharing, customized file syncing and more. You can sign into the Autodesk 360 Cloud from the InfoCenter toolbar using your Autodesk single Sign-In account. If you do not yet have an account, you can create one.

After signing in, your user name is displayed and additional tools are displayed in the drop-down menu including the option to sync your settings with Autodesk 360 Cloud, specify online options, access Autodesk 360 Cloud documents, sign out, and manage account settings.

CREATE AN ACCOUNT

1. Select the **"Sign In"** drop-down menu on the InfoCenter toolbar.

2. Select **"Sign In to Autodesk 360"** from the menu.

The **"Autodesk - Sign In"** dialog box should appear.

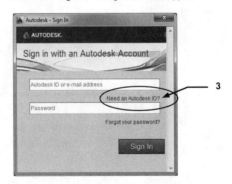

3. Select **"Need an Autodesk ID"**.

Continued on the next page...

HOW TO CREATE AN AUTODESK ACCOUNT....continued

The **"Autodesk - Create Account"** dialog box should appear.

4. Fill in all the required fields.

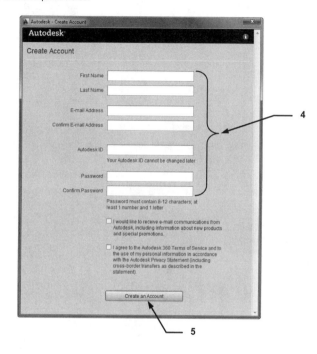

5. Select **"Create an Account"** and then Sign In.

The **first time** you access the Autodesk 360 Cloud, you have the opportunity to specify default Autodesk 360 settings. You may modify these selections later using the **Online ribbon tab** in <u>AutoCAD 2013</u>, and the **Autodesk 360 ribbon tab** in <u>AutoCAD 2014</u>.

HOW TO SAVE A FILE TO AUTODESK 360

1. Select the **Application Menu**.

2. Select **Save As ▶**

3. Select **Drawing to the Cloud**.

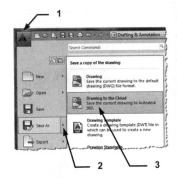

Note: If you are not signed-in to Autodesk 360 the **Sign-In** dialog box will appear.

4. Enter the file name.

5. Select the **Save** button.

Notice "Autodesk 360" Account

Notice "Autodesk 360" directory selected

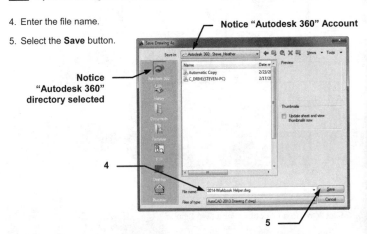

The only difference in the **"Save Drawing As"** dialog box between AutoCAD versions 2013 and 2014 is the **Autodesk 360** directory icon, show below.

AutoCAD 2013

AutoCAD 2014

HOW TO OPEN A FILE FROM AUTODESK 360

1. Select **Open**.

2. Select the **Autodesk 360** directory.

3. Select the **file** to open.

4. Select the **Open** button.

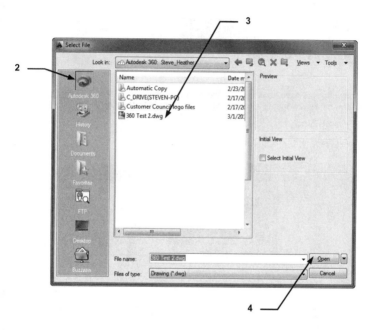

EXITING AUTOCAD

To safely exit AutoCAD follow the instructions below.

1. **Save** all open drawings.

2. Start the **EXIT** procedure using one of the following.

   Ribbon = None
   or
   Application Menu = ⟦ Exit AutoCAD 2014 ⟧
   or
   Keyboard = exit <enter>

If any changes have been made to the drawing since the last **Save As**, the warning box shown below will appear asking if you want to **SAVE THE CHANGES?**

Select **YES**, **NO** or **CANCEL**.

CUSTOMIZING YOUR WHEEL MOUSE

A Wheel mouse has two or more buttons and a small wheel between the two topside buttons. The default functions for the two top buttons and the Wheel are as follows:
Left Hand button is for **input** and can't be reprogrammed.
Right Hand button is for **Enter** or the **shortcut menu**.
The Wheel may be used to <u>Zoom and Pan</u> or <u>Zoom and display</u> the **Object Snap** menu.

The following describes how to select the Wheel functions. After you understand the functions, you may choose to change the setting.
To change the setting you must use the **MBUTTONPAN** variable.

<u>**MBUTTONPAN setting 1:**</u> **(Factory setting)**

ZOOM   Rotate the wheel forward to zoom in.
                 Rotate the wheel backward to zoom out.

ZOOM   Double click the wheel to view entire drawing.
EXTENTS

PAN   Press the wheel and drag the mouse to move the drawing on the screen.

<u>**MBUTTONPAN setting 0:**</u>

ZOOM   Rotate the wheel forward to zoom in.
                 Rotate the wheel backward to zoom out.

OBJECT   Object Snap menu will appear when you press the wheel.
SNAP

<u>**To change the setting:**</u>

1. Type: **mbuttonpan <enter>**
2. Enter **0** or **1** then press **<enter>**

Command Line   **Dynamic Input**

METHODS OF SELECTING OBJECTS

Many AutoCAD commands prompt you to **"select objects"**. This means select the objects that you want the command to effect.

There are 2 methods. **Method 1. Pick**, is very easy and should be used if you have only 1 or 2 objects to select. **Method 2. Window selection**, is a little more difficult but once mastered it is extremely helpful and time saving. Practice the examples shown below.

Method 1. PICK:

First start a command such as ERASE. (Press **E <enter>**) Next you will be prompted to **"Select Objects"**, place the cursor (pick box) on the object but do not press the mouse button yet. The object will highlight. This appearance change is called "Rollover Highlighting". This gives you a preview of which object AutoCAD is recognizing. Press the left mouse button to actually select the highlighted object.

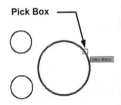

Pick Box

Method 2. WINDOW selection: Crossing and Window

Crossing:
Place your cursor in the area up and to the right of the objects that you wish to select (**P1**) and press the left mouse button. (Do not hold the mouse button down. Just press and release) Then move the cursor down and to the left (**P2**) and press the left mouse button again.

(**Note:** The window will be *green* and outer line is *dashed*.) **Only** the objects that this window **crosses** will be selected.

In the example on the right, all 3 circles have been selected because the Crossing Window crosses a portion of each.

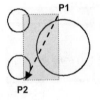

Window:
Place your cursor in the area up and to the left of the objects that you wish to select (**P1**) and press the left mouse button. Then move the cursor down and to the right of the objects (**P2**) and press the left mouse button.

(**Note:** The window will be *blue* and outer line is *solid*.) **Only** the objects that this window **completely enclosed** will be selected.

In the example on the right, only 2 circles have been selected. (The large circle is **not** completely enclosed.)

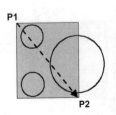

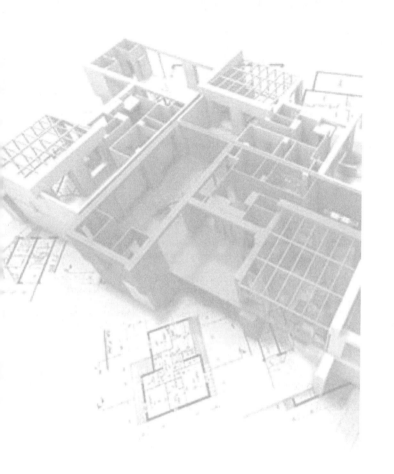

Section 6
Layers

LAYERS

A **LAYER** is like a transparency. Have you ever used an overhead light projector? Remember those transparencies that are laid on top of the light projector? You could stack multiple sheets but the projected image would have the appearance of one document. Layers are basically the same. Multiple layers can be used within one drawing.

The example, on the right, shows 3 layers.
One for annotations (text), one for dimensions
and one for objects.

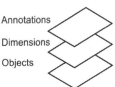

Annotations

Dimensions

Objects

HOW TO USE LAYERS

First you select the layer and then you draw the objects.
Always select the layer first and then draw the objects.

It is good "drawing management" to draw related objects on the same layer.
For example, in an architectural drawing, you would select layer "walls" and then draw the floor plan.
Then you would select the layer "Electrical" and draw the electrical objects.
Then you would select the layer "Plumbing" and draw the plumbing objects.
Each layer can then be controlled independently.
If a layer is <u>Frozen</u>, it is not visible. When you <u>Thaw</u> the layer it becomes <u>visible</u> again.
(Refer to the following pages for detailed instructions for controlling layers.)

HOW TO SELECT A LAYER

1. Go to **Ribbon = Home Tab / Layers Panel**

2. Select the drop down arrow ▼

3. Highlight the desired layer and press the left mouse button.

The selected layer becomes the
"Current" layer. All objects will be
located on this layer until you select
a different layer.

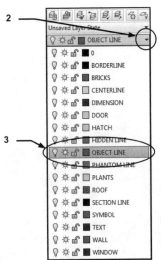

CONTROLLING LAYERS

The following controls can be accessed using the Layer drop down arrow ▼

ON or OFF
If a layer is **ON** it is **visible**. If a layer is **OFF** it is **not visible**.
Only layers that are **ON** can be edited or plotted.

FREEZE or THAW
Freeze and **Thaw** are very similar to On and Off. A Frozen layer is **not visible** and a Thawed layer is **visible**. Only thawed layers can be edited or plotted.

Additionally:
a. Objects on a Frozen layer **cannot** be accidentally erased
b. When working with large and complex drawings, freezing saves time because frozen layers are not **regenerated** when you zoom in and out.

LOCK or UNLOCK
Locked layers are visible but cannot be edited.
They are visible so they **will** be plotted.

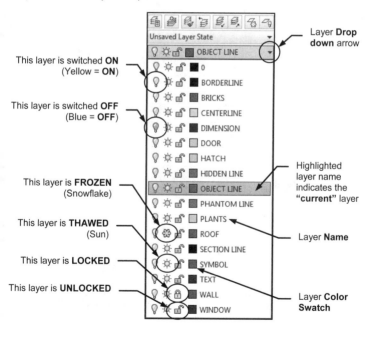

Layer **Drop down** arrow

This layer is switched **ON**
(Yellow = **ON**)

This layer is switched **OFF**
(Blue = **OFF**)

This layer is **FROZEN**
(Snowflake)

This layer is **THAWED**
(Sun)

This layer is **LOCKED**

This layer is **UNLOCKED**

Highlighted layer name indicates the **"current"** layer

Layer **Name**

Layer **Color Swatch**

CONTROLLING LAYERS....continued

To access the following options you must use the **Layer Properties Manager**.
You may also access the options listed on the previous page within this dialog box.

To open the **Layer Properties Manager** use one of the following:

Ribbon = Home Tab / Layers Panel /
or
Keyboard = LA <enter>

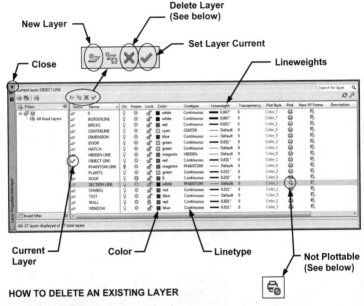

HOW TO DELETE AN EXISTING LAYER

1. Highlight the layer name to be deleted.
2. Select the **Delete Layer** tool.
   or
1. Highlight the layer name to be deleted.
2. Right click and select **Delete Layer**.

PLOT or NOT PLOTTABLE
This tool prevents a layer from plotting even though it is visible within the Drawing Area.
A **Not Plottable** layer will not be displayed when using **Plot Preview**.
If the Plot tool has a slash the layer will not plot.

LAYER COLOR

Color is not merely to make a pretty display on the screen. Layer colors can help define objects. For example, you may assign color Green for all doors. Then, at a glance, you could identify the door and the layer by their color.

Here are some additional things to consider when selecting the colors for your layers.

Consider how the colors will appear on the paper.
(Pastels do not display well on white paper.)

Consider how the colors will appear on the screen.
(Yellow appears well on a black background but not on white.)

How to change the color of a layer.

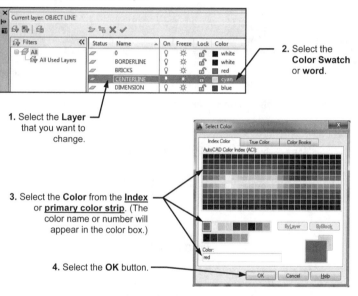

2. Select the **Color Swatch** or **word**.

1. Select the **Layer** that you want to change.

3. Select the **Color** from the __Index__ or __primary color strip__. (The color name or number will appear in the color box.)

4. Select the **OK** button.

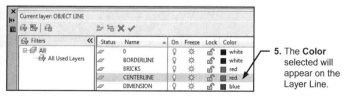

5. The **Color** selected will appear on the Layer Line.

LINEWEIGHTS

A **Lineweight** means **"how heavy or thin is the object line"**.

It is **"good drawing management"** to establish a contrast in the lineweights of entities.

In the example below the rectangle has a heavier lineweight than the dimensions. The contrast in lineweights makes it easier to distinguish between entities.

LINEWEIGHT SETTINGS
Lineweights are plotted with the exact width of the lineweight assigned.
But you may **adjust** how they are **displayed on the screen**. (Refer to **#4** below)

IMPORTANT: Before assigning lineweights you should first select the **Units for Listing** and **Adjust Display Scale** as shown below.

1. Select the **Lineweight Settings** dialog box using one of the following:

    Keyboard = LW <enter>
    or
    Status Bar = Right-Click on the LWT button and select Settings

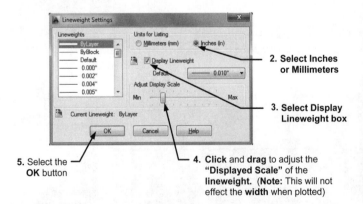

5. Select the **OK** button

2. **Select Inches or Millimeters**

3. **Select Display Lineweight box**

4. **Click** and **drag** to adjust the **"Displayed Scale"** of the **lineweight**. (**Note:** This will not effect the **width** when plotted)

<u>NOTE:</u> The **Lineweight settings** will be **saved to the computer not the drawing** and will remain until you change them.

ASSIGNING LINEWEIGHTS

Note: Before assigning **Lineweights** to Layers make sure your **Lineweight settings** (Units for listing and Adjust Display scale) are correct. Refer to the previous page.

ASSIGNING LINEWEIGHTS TO LAYERS

1. Select the **Layer Properties Manager** using one of the following:

Ribbon = Home Tab / Layers Panel /
or
Keyboard = LA <enter>

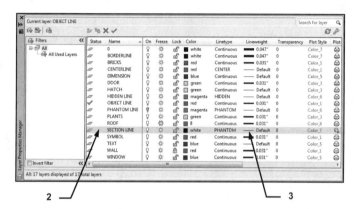

2. Highlight a Layer (Click on the name).

3. Click on the Lineweight for that layer.

4. Scroll and select a Lineweight from the list.

5. Select the **OK** button.

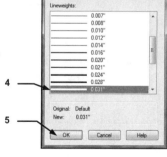

Note:
Lineweight selections will be saved within the **current** drawing and will not effect any other drawing.

CREATING NEW LAYERS

Using layers is an important part of managing and controlling your drawing. It is better to have too many layers than too few. You should draw like objects on the same layer. For example, place all doors on the layer "door" or centerlines on the layer "centerline".

When you create a new layer you will assign a **name**, **color**, **linetype**, **lineweight**, and **transparency** and whether or not it should plot.

1. Select the **Layer Properties Manager** using one of the following:

Ribbon = Home Tab / Layers Panel /
or
Keyboard = LA <enter>

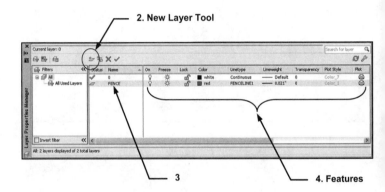

2. New Layer Tool

3

4. Features

2. Select the **New Layer Tool** and a new Layer will appear.

3. Type the new **Layer Name** and press **<enter>**

4. Select any of the **features** and a dialog box will appear.

Features:
Refer to the previous pages for controlling and selecting Color and Lineweights.
Refer to the next few pages for Linetypes and Transparency.

LOADING AND SELECTING LAYER LINETYPES

In an effort to conserve data within a drawing file, AutoCAD automatically loads only one linetype called **"Continuous"**. If you would like to use other linetypes, such as "dashed" or "fenceline", you must **Load** them into the drawing as follows:

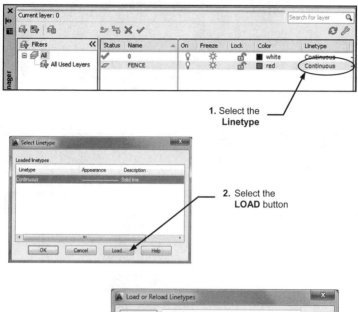

1. Select the **Linetype**

2. Select the **LOAD** button

3. Select a **Linetype**

4. Select the **OK** button

Continued on the next page...

LOADING AND SELECTING LAYER LINETYPES...continued

5. Select the **Linetype** to assign to the **Layer**

6. Select the **OK** button

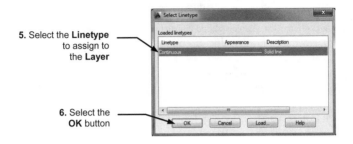

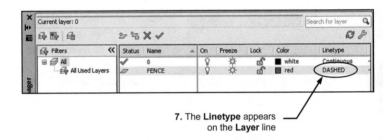

7. The **Linetype** appears on the **Layer** line

LAYER MATCH

If you draw an object on the wrong layer you can easily change it to the desired layer using the **Layer Match** command. You first select the object that needs to be changed and then select an object that is on the correct layer (object on destination layer).

1. Select the **Layer Match** command using one of the following:

   Ribbon = Home Tab / Layers Panel /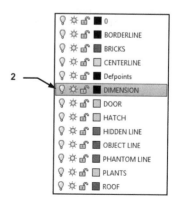
   or
   Keyboard = laymch <enter>

   Command: _laymch
2. Select objects to be changed: *select the objects that need to be changed.*

3. Select objects: *select more objects or <enter> to stop selecting.*

4. Select object on destination layer or [Name]: *select the object that is on the layer*
   that you want to change to.

Note:
You may also easily change the layer of an object to another layer as follows:

1. Select the object that you wish to change.

2. Select the Layer drop down menu and select the layer that you wish the object to be placed on.

For example:
If you had a dimension on the **OBJECT LINE** layer by mistake.

1. Select the dimension that is mistakenly on the **OBJECT LINE** layer.

2. Select the **DIMENSION** layer from the drop-down menu.

LAYER TRANSPARENCY

Each layer may be assigned a transparency percentage from 0 to 90 percent.
0 would not be transparent at all and 90 would be 90% transparent.

ASSIGNING TRANSPARENCY TO LAYERS

1. Select the **Layer Properties Manager** using one of the following:

   Ribbon = Home Tab / Layers Panel /
   or
   Keyboard = LA <enter>

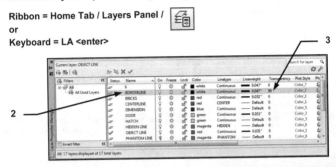

2. Highlight a **Layer** (Click on the name)
3. Click on **Transparency** for that layer.
4. Select a **Transparency** from the list.
5. Select the **OK** button.

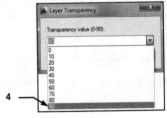

Controlling Transparent display
You may toggle the display of Transparent objects **ON** or **OFF** by selecting the **TPY** button on the Status bar.

Transparency ON   Transparency OFF

Note: Transparency selections will be saved within the **current** drawing and will not effect any other drawing.

Plotting Transparent Objects
Plotting transparency is disabled by default. To plot transparent objects, check the Plot transparency option in either the Plot dialog box or the Page Setup dialog box.

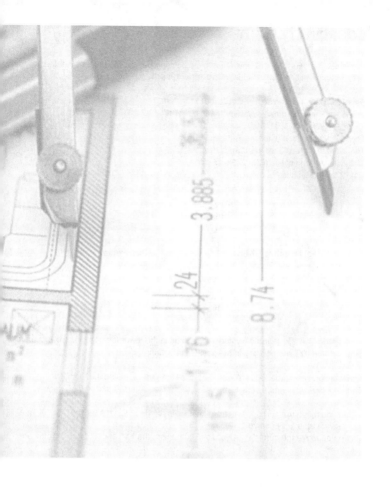

Section 7
Input Options

COORDINATE INPUT

In this section you will learn how to place objects in **specific locations** by entering coordinates. This process is called **Coordinate Input**.

AutoCAD uses the *Cartesian Coordinate System*.
The Cartesian Coordinate System has 3 axes, **"X"**, **"Y"** and **"Z"**.

The **X** is the Horizontal axis. (Right and Left)
The **Y** is the Vertical axis. (Up and Down)
The **Z** is Perpendicular to the X and Y plane.
(The **Z** axis, which is not discussed in this book, is used for 3-Dimensional drawing.)

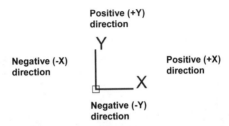

Look at the User Coordinate System (UCS) icon in the lower left corner of your screen. The **"X"** and **"Y"** are pointing in the positive direction.

The location where the **"X"**, **"Y"** and **"Z"** axes intersect is called the **ORIGIN**. *The **Origin** always has a coordinate value of X=0, Y=0 and Z=0 (0,0,0)*

When you move the <u>cursor</u> away from the Origin, in the positive direction, the **"X"** and **"Y"** coordinates are positive.
When you move the cursor in the opposite direction, the **"X"** and **"Y"** coordinates are negative.

Using this system, every point on the screen can be specified using positive or negative **"X"** and **"Y"** coordinates.

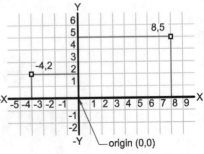

ABSOLUTE COORDINATES

There are **3** types of Coordinate input, **Absolute, Relative** and **Polar**.

ABSOLUTE COORDINATES

When inputting absolute coordinates the input format is: **"X", "Y"**   (that is: X comma Y)

Absolute coordinates come *from the ORIGIN* and are typed as follows: **8, 5**

The first number (8) represents the **X-axis** (horizontal) distance from the **Origin**, and the second number (5) represents the **Y-axis** (vertical) distance from the **Origin**. The two numbers must be separated by a **comma**.

An absolute coordinate of **4, 2** will be **4** units to the right (horizontal) and **2** units up (vertical) from the current location of the **Origin**.

An absolute coordinate of **-4, -2** will be **4** units to the left (horizontal) and **2** units down (vertical) from the current location of the **Origin**.

The following are examples of Absolute Coordinate input. Notice where the **Origin** is located in each example.

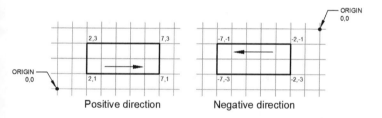

Positive direction   Negative direction

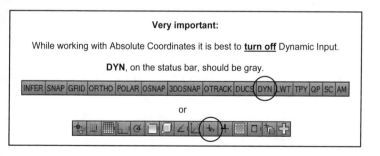

Very important:

While working with Absolute Coordinates it is best to **turn off** Dynamic Input.

DYN, on the status bar, should be gray.

or

RELATIVE COORDINATES

RELATIVE COORDINATES

Relative coordinates come *from the last point entered*. (Not from the Origin)

The first number represents the **X-axis** (horizontal) and the second number represents the **Y-axis** (vertical) just like the absolute coordinates.

To distinguish the relative coordinates from absolute coordinates the two numbers must be preceded by an **@** symbol in addition to being separated by a **comma**.

A Relative coordinate of **@5, 2** will go to the right **5** units and up **2** units from the **last point entered**.

A Relative coordinate of **@-5, -2** will go to the left **5** units and down **2** units from the **last point entered**.

The following is an example of Relative Coordinate input:

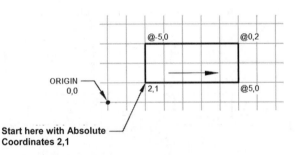

Start here with Absolute Coordinates 2,1

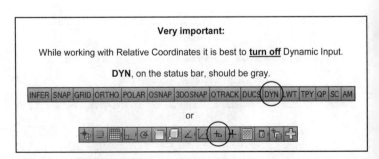

Very important:

While working with Relative Coordinates it is best to **turn off** Dynamic Input.

DYN, on the status bar, should be gray.

EXAMPLES OF COORDINATE INPUT

Scenario 1.
If you want to draw a line with the first endpoint "at the Origin" and the second endpoint
3 units in the positive **X** direction.

0,0 ———————— 3,0

1. Select the **Line** command.
2. You are prompted for the first endpoint: *Type 0, 0 <enter>*
3. You are then prompted for the second endpoint: *Type 3, 0 <enter>*

What did you do?
The first endpoint coordinate input, **0, 0** means that you do not want to move away from
the Origin. You want to start **"ON"** the Origin.

The second endpoint coordinate input, **3, 0** means that you want to move **3** units in the
positive **X** axis. The **"0"** means you do not want to move in the **Y axis**. So the line will
be exactly horizontal.

Scenario 2.
You want to start a line **1** unit to the right of the origin and **1** unit above, and the line will
be **4** units in length, perfectly vertical.

@0,4

1. Select the **Line** command.
2. You are prompted for the first endpoint: *Type 1, 1 <enter>*
3. You are prompted for the second endpoint: *Type @0, 4 <enter>*

1,1

What did you do?
The first endpoint coordinate input, **1, 1** means you want to move **1** unit in the **X axis**
direction and **1** unit in the **Y axis** direction.

The second endpoint coordinate input **@0, 4** means you do not want to move in the **X
axis** "from the last point entered" but you do want to move in the **Y axis** "from the last
point entered. (Remember the **@** symbol is only necessary if you are not using **DYN**)

Scenario 3.
Now try drawing **5** connecting line segments.
(Watch for the negatives)
1. Select the **Line** command.
2. First endpoint: *2, 4 <enter>*
3. Second endpoint: *@ 2, -3 <enter>*
4. Second endpoint: *@ 0, -1 <enter>*
5. Second endpoint: *@ -1, 0 <enter>*
6. Second endpoint: *@ -2, 2 <enter>*
7. Second endpoint: *@ 0, 2 <enter> <enter>*

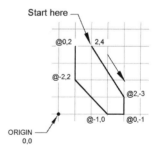

Note: If you enter an incorrect coordinate, just hold down the **Shift key** and press **U**
then **<enter>**, the last segment will disappear and you will have another chance at
entering the correct coordinate.

DIRECT DISTANCE ENTRY (DDE)

DIRECT DISTANCE ENTRY is a combination of keyboard entry and cursor movement. **DDE** is used to specify distances in the horizontal or vertical axes from the **last point entered**. DDE is a **Relative Input**. Since it is used for Horizontal and Vertical movements, **Ortho must be ON**.

(**Note:** to specify distances on an angle, refer to **Polar Input**)

Using DDE is simple. Just move the cursor and type the distance.
Negative and positive is understood automatically by moving the cursor up (positive), down (negative), right (positive) or left (negative) from the last point entered. No minus or @ sign necessary.

Moving the cursor to the **right** and typing **5 <enter>** tells AutoCAD that the **5** is positive and Horizontal.
Moving the cursor to the **left** and typing **5 <enter>** tells AutoCAD that the **5** is negative and Horizontal.
Moving the cursor **up** and typing **5 <enter>** tells AutoCAD that the **5** is positive and Vertical.
Moving the cursor **down** and typing **5 <enter>** tells AutoCAD that the **5** is negative and Vertical.

EXAMPLE:

1. **Ortho** must be **ON**. Grid OFF
2. Select the **Line** command.
3. Type: *1, 2 <enter>* to enter the first endpoint using Absolute coordinates.
4. Now move your cursor to the **right** and type: *5 <enter>*
5. Now move your cursor **up** and type: *4 <enter>*
6. Now move your cursor to the **left** and type: *5 <enter> <enter> to stop*

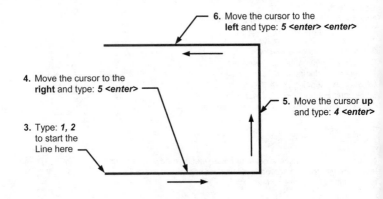

6. Move the cursor to the **left** and type: *5 <enter> <enter>*

4. Move the cursor to the **right** and type: *5 <enter>*

5. Move the cursor **up** and type: *4 <enter>*

3. Type: *1, 2* to start the Line here

POLAR COORDINATE INPUT

Previously you learned to control the length and direction of horizontal and vertical lines using Relative Input and Direct Distance Entry. Now you will learn how to control the length and **ANGLE** of a line using **POLAR** Coordinate Input..

UNDERSTANDING THE *"POLAR DEGREE CLOCK"*

Previously when drawing Horizontal and Vertical lines you controlled the direction using a **Positive** or **Negative** input. *Polar Input is different*. The Angle of the line will determine the direction.

For example: If I want to draw a line at a 45 degree angle towards the upper right corner, you would use the angle 45. But if I want to draw a line at a 45 degree angle towards the lower left corner, you would use the angle 225.

You may also use Polar Input for Horizontal and Vertical lines using the angles 0, 90, 180 and 270. No negative input is required.

POLAR DEGREE CLOCK

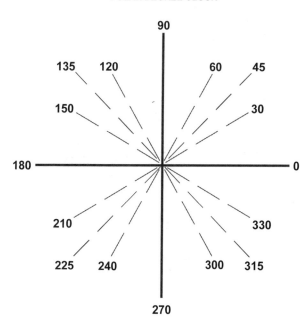

POLAR COORDINATE INPUT....continued

DRAWING WITH POLAR COORDINATE INPUT

When entering polar coordinates the first number represents the **Distance** and the second number represents the **Angle**. The two numbers are separated by the **less than (<)** symbol. The input format is: **distance < angle**

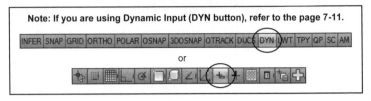

Note: If you are using Dynamic Input (DYN button), refer to the page 7-11.

A Polar coordinate of **@6<45** will be a distance of **6 units** and an angle of **45 degrees** from the **last point entered**.

Here is an example of Polar input for 4 line segments.

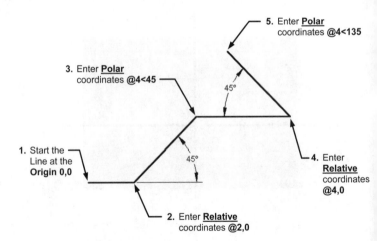

POLAR TRACKING

Polar Tracking can be used instead of **Dynamic Input**. When Polar Tracking is *"ON"*, a dotted *"tracking"* line and a *"tool tip"* box appear. The tracking line.... *"snaps"* to a **preset angle increment** when the cursor approaches one of the preset angles. The word *"Polar"*, followed by the *"distance"* and *"angle"* from the last point appears in the box.

← Tracking Line

← Polar Tool Tip box

HOW TO SET THE INCREMENT ANGLE

1. Right Click on the **POLAR** button on the Status Bar and select **"SETTINGS"**, or select an **angle from the list**.

POLAR ANGLE SETTINGS

Increment Angle: Choose from the Increment Angle list including 90, 45, 30, 22.5, 18, 15, 10 and 5. It will also snap to the selected angles multiples. For example: if you choose 30 it will snap to 30, 60, 90, 120, 150, 180, 210, 240, 270, 300, 330 and 0.

Additional Angles: Check this box if you would like to use an angle other than one in the Incremental Angle list. For example: 12.5

New: You may add an angle by selecting the **"New"** button. You will be able to snap to this new angle in addition to the incremental Angle selected. <u>But you will not be able to snap to it's multiple</u>. For example, if you selected 7, you would not be able to snap to 14

Delete: Deletes an Additional Angle. Select the Additional angle to be deleted and then the **"Delete"** button.

POLAR ANGLE MEASUREMENT

Absolute: Polar tracking **angles** are relative to the **UCS**.

Relative to last segment: Polar tracking **angles** are relative to the **last segment**.

POLAR SNAP

Polar Snap is used with **Polar Tracking** to make the cursor snap to specific *distances* and *angles*. If you set Polar Snap distance to 1 and Polar Tracking to angle 30 you can draw lines 1, 2, 3 or 4 units long at an angle of 30, 60, 90 etc. without typing anything. You just move the cursor and watch the tool tips.

SETTING THE POLAR SNAP

1. Set the **Polar Tracking Increment Angle**.

2. Right Click on the **SNAP** button on the Status Bar and select **"SETTINGS"**.

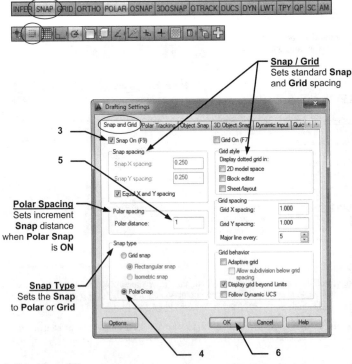

Snap / Grid
Sets standard **Snap** and **Grid** spacing

Polar Spacing
Sets increment **Snap** distance when **Polar Snap** is ON

Snap Type
Sets the **Snap** to **Polar** or **Grid**

3. Select **Snap ON**.

4. Select **PolarSnap**.

5. Set the **Polar Distance**.

6. Select the **OK** button.

DYNAMIC INPUT

To help you keep your focus in the "drawing area", AutoCAD has provided a command interface called **Dynamic Input**. You may input information within the Dynamic Input tool tip box instead of on the command line.

When AutoCAD prompts you for the **First point** the Dynamic Input tool tip displays the **Absolute: X, Y** distance from the Origin.
Enter the **X** dimension, <u>press the Tab key</u>, enter the **Y** dimension then **<enter>**.

First Point

When AutoCAD prompts you for the **Second** and all **Next points** the Dynamic Input tool tip displays the **Relative: Distance and Angle** from the last point entered.
Enter the **distance**, <u>press the Tab key</u>, move the cursor in the approximate desired angle and enter the **angle** then **<enter>**. (**Note:** The @ is not necessary)

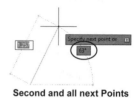

Second and all next Points

How to turn Dynamic Input ON or OFF

Select the **DYN** button on the status bar or use the **F12** key.

Here is an example of Dynamic input for 4 line segments.

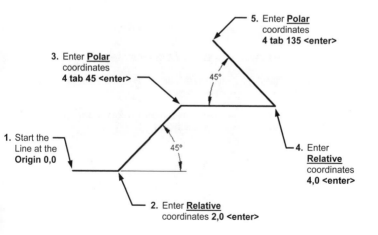

5. Enter **Polar** coordinates **4 tab 135 <enter>**

3. Enter **Polar** coordinates **4 tab 45 <enter>**

1. Start the Line at the **Origin 0,0**

4. Enter **Relative** coordinates **4,0 <enter>**

2. Enter **Relative** coordinates **2,0 <enter>**

DYNAMIC INPUT....continued

To enter Cartesian coordinates (X and Y)

1. Enter an **"X"** coordinate value and a __comma__.

2. Enter an **"Y"** coordinate value **<enter>**.

To enter Polar coordinates (from the last point entered)

1. Enter the **distance** value from the last
   point entered.

2. Press the **Tab** key.

3. Move the cursor in the approximate
   direction and enter the **angle** value **<enter>**.

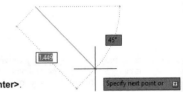

Note: Move the cursor in the approximate direction
and enter an angle value of **0-180** only.
Dynamic Input does not use 360 degrees.

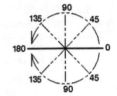

How to specify Absolute or Relative coordinates while using Dynamic Input.

To enter **absolute** coordinates when relative coordinate format is displayed in the
tooltip. Enter # to temporarily override the setting.

To enter **relative** coordinates when absolute coordinate format is displayed in the
tooltip. Enter @ to temporarily override the setting.

Note about Ortho

You may toggle Ortho **ON** and **OFF** by holding down the **shift** key.
This is an easy method to use **D**irect **D**istance **E**ntry while using Dynamic Input.

Section 8
Miscellaneous

BACKGROUND MASK

Background Mask inserts an opaque background so that objects under the text are covered. (masked) The mask will be a rectangular shape and the size will be controlled by the **"Border offset factor"**. The color can be the same as the drawing background or you may select a different color.

1. Select the **Multiline Text** command.

2. **Right click** in the Text Area and select **"Background Mask"** from the menu.

3. The **Background Mask** dialog box appears.

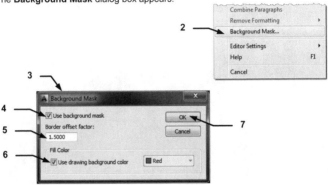

4. Turn this option **ON** by selecting the **"Use background Mask"** box.

5. Enter a value for the **"Border offset factor"**. The value is a factor of the text height. 1.0 will be exactly the same size as the text. 1.5 (the default) extends the background by 0.5 times the text height. The width will be the same width that you defined for the entire paragraph.

6. In the **Fill Color** area, select the **"Use drawing background color"** box to make the background the same color as the drawing background. To specify a color, uncheck this box and select a color. (Note: if you use a color you may need to adjust the "draworder" using Tools / Draworder to bring the text to the front.)

7. Select **OK** when all settings are complete.

No Mask

Mask using the drawing background color

Mask using a different color

BACK UP FILES

When you save a drawing file, Autocad creates a file with a **".dwg"** extension.
For example, if you save a drawing as **"12b"**, Autocad saves it as **"12b.dwg"**.
The next time you save that same drawing, Autocad replaces the old with the new and renames the old version **"12b.bak"**. The old version is now a **back up** file.
(Only 1 backup file for each drawing file is stored.)

How to open a back up file:

You can't open a **".bak"** file.
It must first be renamed with a **".dwg"** file extension.

How to view the list of back up files:

The backup files will be saved in the same location as the drawing file.
You must use Windows Explorer to locate the **".bak"** files.

How to rename a back up file:

1. **Right click** on the drawing file name.
2. Select **"Rename"**.
3. Change the **".bak"** extension to **".dwg"** and press **<enter>**.

RECOVERING A DRAWING

In the event of a program failure or a power failure any open files should be saved automatically.

When you attempt to re-open the drawing the **Drawing Recovery Manager** will display a list of all drawing files that were open at the time of a program or system failure. You can preview and open each **".dwg"** or **".bak"** file to choose which one should be saved as the primary file.

GRIPS

Grips are little boxes that appear if you select an object when no command is in use. Grips can be used to quickly edit objects. You can copy, lengthen, mirror, move, rotate, scale, and stretch objects using grips.

The following is a brief overview on how to use three of the most frequently used options. Grips have many more options and if you like the examples below, you should research them further in the AutoCAD help menu.

1. Select the object (no command can be in use while using grips).
2. Select one of the **blue** grips. It will turn to **"red"**. This indicates that it is **"hot"**. The **"Hot"** grip is the **basepoint**.
3. The editing modes will be displayed on the command line. You may cycle through these modes by pressing the SPACEBAR or ENTER key or use the shortcut menu.
4. After editing you must press the **ESC** key to deactivate the grips on that object.

SELECTING A GRIP:

When you select a grip it becomes **"HOT"**.

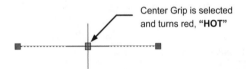

Center Grip is selected
and turns red, **"HOT"**

MOVING AN OBJECT:

1. Select the object.

2. Left click to select the grip in the middle of the object.

3. Move the cursor to the new location.

4. Left click again to apply the new location.

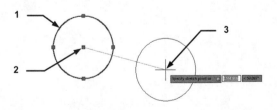

Continued on the next page...

GRIPS....continued

COPYING AN OBJECT:

1. Select the object.

2. Select the grip.

3. Select the **COPY** option from the Dynamic Input or Command Line options.

4. Move the cursor to the new location for the copy(s) and left click.

Note: Grips will allow you to continue making copies until you press the **ENTER** or **ESC** keys to stop.

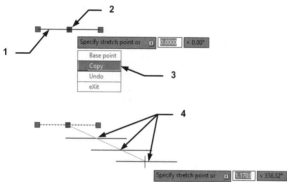

LENGTHEN AN OBJECT:

1. Select the object.

2. Hold the cursor over a grip.

3. Select the **Lengthen** option.

4. **Move** the cursor to lengthen the object, then **left click**. Or **type** in the desired size and then press **<enter>**.

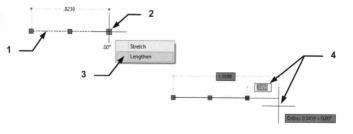

OBJECT SNAP

Object Snap enables you to **"Snap"** to **"Objects"** in very specific and accurate locations on the objects. For example; you could snap to the endpoint of a Line using the **"Endpoint"** snap, or you could snap to the center of a Circle using the **"Center"** snap.

<u>Note:</u> You must select a command such as LINE before you can use Object Snaps.

How to select from the Object Snap Menu

1. You must select a command, such as LINE, before you can select Object Snap.

2. While holding down the **shift key, press the right mouse button**. The menu shown below should appear.

3. **Highlight** the required snap and **press the left mouse button** to select.

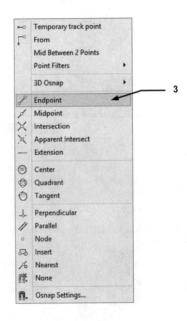

The following object snaps will be discussed on the next page:

Endpoint, Midpoint, Intersection, Center, Quadrant, Tangent, and Perpendicular.

OBJECT SNAP....continued

Object Snap Definitions:

Object snap is used when AutoCAD prompts you to place an object. Object snap allows you to place objects very accurately.

A step by step example of "How to use object snap" is shown on the next page.

<u>Note:</u> You may type the **3 bold letters** shown rather than select from the menu.

 ENDpoint

Snaps to the closest endpoint of a Line, Arc or polygon segment. Place the cursor on the object close to the end and the cursor will snap like a magnet to the end of the line.

 MIDpoint

Snaps to the middle of a Line, Arc or Polygon segment. Place the cursor anywhere on the object and the cursor will snap like a magnet to the midpoint of the line.

 INTersection

Snaps to the intersections of any two objects. Place the cursor directly on top of the intersection or select one object and then the other and Autocad will locate the intersection.

 CENter

Snaps to the center of an Arc, Circle or Donut. Place the cursor on the object, or at the approximate center location and the cursor will snap like a magnet to the center.

 QUAdrant

Snaps to a 12:00, 3:00, 6:00 or 9:00 o'clock location on a circle or ellipse. Place the cursor on the circle near the desired quadrant location and the cursor will snap to the closest quadrant.

 TANgent

Calculates the tangent point of an Arc or Circle. Place the cursor on the object as near as possible to the expected tangent point.

 PERpendicular

Snaps to a point perpendicular to the object selected. Place the cursor anywhere on the object then pull the cursor away from object and press the left mouse button.

HOW TO USE OBJECT SNAP

The following is an example of attaching a line segment to previously drawn vertical lines. The new line will start from the upper endpoint, to the midpoint, to the lower endpoint.

1. Turn Off **SNAP**, **ORTHO** and **OSNAP** on the <u>Status Bar</u>. (Gray is OFF)

2. Select the **Line** command.

3. Draw two vertical lines as shown below (they don't have to be perfectly straight).

4. Select the **Line** command again.

5. Hold the **shift key down** and press the **right mouse button**.

6. Select the Object snap **"Endpoint"** from the object snap menu.

7. Place the cursor close to the upper endpoint of the left hand line.

   The cursor should snap to the end of the line like a magnet. A little square and an "Endpoint" tooltip are displayed.

8. Press the **left mouse button** to attach the new line to the upper endpoint of the previously drawn vertical line. (**Do not end the Line command yet.**)

Continued on the next page...

HOW TO USE OBJECT SNAP....continued

9. Now hold the **shift key down** and **press the right mouse button** and select the **"Midpoint"** object snap option.

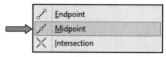

10. Move the cursor to approximately the middle of the right hand vertical line.

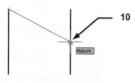

The cursor should snap to the midpoint of the line like a magnet.
A little triangle with a "Midpoint" tool tip are displayed.

11. Press the **left mouse button** to <u>attach</u> the new line to the midpoint of the previously drawn vertical line. (**Do not end the Line command yet.**)

12. Now hold the **shift key down** and press the **right mouse button** and select the object snap **"Endpoint"** again.

13. Move the cursor close to the lower endpoint of the left hand vertical line.

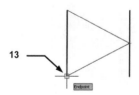

The cursor should snap to the end of the line like a magnet.
A little square and a tooltip are displayed.

14. Press the **left mouse button** to <u>attach</u> the new line to the endpoint of the previously drawn vertical line.

15. Stop the Line command and disconnect by pressing **<enter>**.

RUNNING OBJECT SNAP

Selecting Object Snap is not difficult but AutoCAD has provided you with an additional method to increase your efficiency by allowing you to preset frequently used object snap options. This method is called **RUNNING OBJECT SNAP**.

When **Running Object Snap** is active the cursor will automatically snap to any preset object snap locations thus eliminating the necessity of invoking the object snap menu for each locations.

First you must **set the running object snaps**, and second, you must **turn ON the Running Object Snap option**.

SETTING RUNNING OBJECT SNAP:

1. Select the **Running Object Snap** dialog box using one of the following:

   Keyboard = DS <enter>
   or
   Status Bar = Right Click on the OSNAP button and select "SETTINGS"

2. Select the **Object Snap** tab.

3. Select the Object Snaps desired.

   (In the example below object snap **Endpoint**, **Midpoint** and **Intersection** have been selected).

Note:

Try not to select more than 3 or 4 running object snaps at a time.

If you select too many snaps the cursor will flit around trying to snap to multiple snap locations. And possibly snap to the wrong location.

You will lose control and it will confuse you.

4. Select the **OK** button.

5. Turn **ON** the **OSNAP** button on the Status Bar. (Blue is ON)

PAN

After you zoom in and out or adjust the scale of a viewport the drawing within the viewport frame may not be positioned as you would like it. This is where **PAN** comes in handy. **PAN** will allow you to move the drawing around, within the viewport, without affecting the size or scale.

Note: Do not use the **MOVE** command. You do not want to actually move the original drawing. You only want to slide the viewport image, of the original drawing, around within the viewport.

How to use the PAN command.

1. Select a layout tab (paper space).

2. **Unlock** the viewport if it is locked.

3. Click inside a **viewport**.

4. Select the **PAN** command using one of the following:

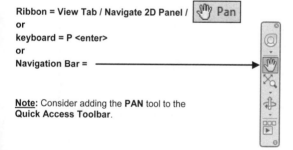

    Ribbon = View Tab / Navigate 2D Panel /   🖐 Pan
    or
    keyboard = P <enter>
    or
    Navigation Bar = ⎯⎯⎯⎯⎯⎯⎯⎯⎯⎯→

    Note: Consider adding the **PAN** tool to the
    Quick Access Toolbar.

5. Place the cursor inside the viewport and <u>hold the left mouse button</u> down while moving the cursor. (Click and drag) When the drawing is in the desired location release the mouse button.

6. Press the **Esc** key or press **<enter>** to end the **PAN** command.

7. **Lock** the viewport.

Refer to the Examples on the next page.

Continued on the next page...

PAN....continued

Before PAN

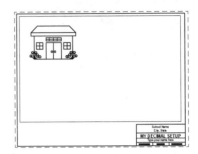

**Double Click in the
Viewport to activate it.**

**Use PAN
(Click, Drag, Release)**

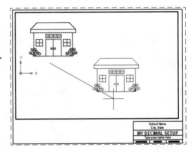

After PAN

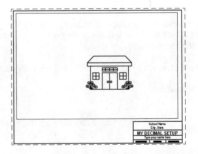

PROPERTIES PALETTE

The **Properties Palette**, shown below, makes it possible to change an object's properties. You simply open the Properties Palette, select an object and you can change any of the properties that are listed.

How to open the Properties Palette:

Ribbon = Home Tab / Properties Panel / ↘
or
Keyboard = Ctrl + 1

(An example of how to use the Properties Palette is on the next page.)

Close Palette
To close the Palette click on
The **"X"** or press **Ctrl + 1**

Auto-Hide feature
(Click On and Off here)
The Palette will collapse into
a vertical bar if **ON**. It will
reactivate if you pass the
cursor over the gray bar

Multiple selected objects list
If you have selected multiple
objects, select individually from
this drop down list

Move
To move, click and drag the
gray title bar to a new location

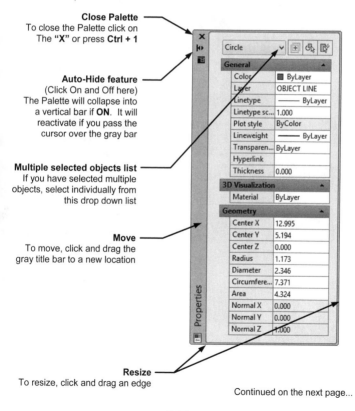

Resize
To resize, click and drag an edge

Continued on the next page...

PROPERTIES PALETTE....continued

Example of editing an object using the Properties Palette:

1. Draw a **2.000** Radius circle.

2. Open the **Properties Palette** and select the **Circle**. (The Properties for the Circle should appear. You may change any of the properties listed in the Properties Palette for this object. When you press **<enter>** the circle will change.)

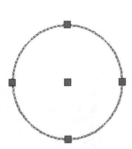

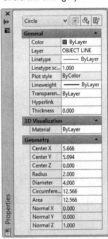

3. Highlight and change the **"Radius"** to **1.000** and the **"Layer"** to HIDDEN LINE <enter>.

   The Circle got smaller and the Layer changed as shown below.

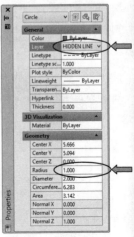

QUICK PROPERTIES PANEL

The **Quick Properties Panel**, shown below, will only appear if you have it set to **ON**. The Quick Properties Panel displays fewer properties and appears when you click once on an object. You may make changes to the objects properties using Quick Properties just as you would using the Properties Palette. (AutoCAD is just giving you another option)

How to turn Quick Properties Panel On or OFF:

1. Select and turn **ON** the **QP** button on the status bar.

2. Select an object.

3. The **Quick Properties Panel** appears.

Note: The list depends on the type of object you have selected, such as; Circle, Line, Rectangle, etc.

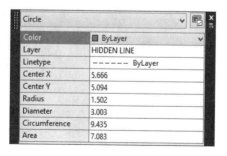

Circle	
Color	■ ByLayer
Layer	HIDDEN LINE
Linetype	– – – – – ByLayer
Center X	5.666
Center Y	5.094
Radius	1.502
Diameter	3.003
Circumference	9.435
Area	7.083

Refer to the next page to **"Customize"** the **Quick Properties Panel**.

CUSTOMIZING THE QUICK PROPERTIES PANEL

You may add or remove properties from the **Quick Properties Panel**. <u>And it is easy</u>.

1. Select the **Customize** button.

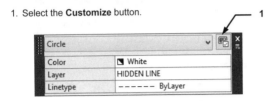

2. Select the Object, from the list, that you would like to customize.

3. **Check** the boxes for the properties that you **want to appear**.
   Uncheck the boxes that you **do not want to appear**.

4. Select **Apply** then **OK**.

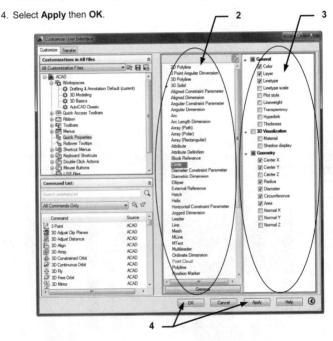

<u>Note:</u> The customizing is saved to the computer not the drawing file.

COMMAND LINE ENHANCEMENTS

(**Note:** Not available in AutoCAD 2013)

The Command Line in AutoCAD 2014 has been extensively modified to further assist the user in searching for commands.

AUTOCORRECT

If you mistyped a command in previous versions of AutoCAD, the system would respond with **"Unknown command"**, 2014 will now **AutoCorrect** to the most relevant command.

In the example below, if you entered **CIRKLE**, the system will respond with **CIRCLE**, and any other commands that contain the word **CIRCLE**.

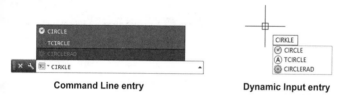

Command Line entry   **Dynamic Input entry**

AutoCAD 2014 also has a new **AutoCorrect List** which is stored in the system, if you mistype a command three times or more, that mistyped command will be stored in the **AutoCorrect List** along with the correct spelling of the command.

You can access the AutoCorrect List by selecting:

Ribbon = Manage Tab / Customization Panel / Edit Aliases ▾ / Edit AutoCorrect List

Continued on the next page...

An example of the **AutoCorrect List** is shown below with two commands that have been mistyped, and with their correct spelling.

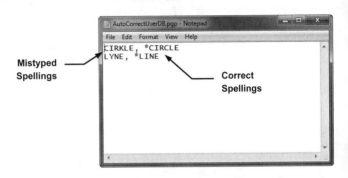

Mistyped Spellings

Correct Spellings

AUTOCOMPLETE

The **AutoComplete** in AutoCAD has been further enhanced and now supports mid-string searches. In previous versions the AutoComplete only displayed command suggestions beginning with the word you entered, it will now display command suggestions with the word you enter, anywhere within it.

In the example below, if you enter **SETTINGS**, AutoComplete will respond with various suggestions with the word **SETTINGS** anywhere within a command.

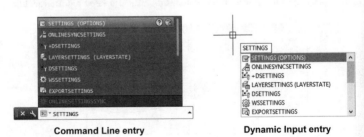

Command Line entry   **Dynamic Input entry**

ADAPTIVE SUGGESTIONS

When you first use AutoCAD 2014, commands in the suggestion list are displayed in the order of usage which is based on general customer data. As you use AutoCAD more and more, the commands will be displayed according to your usage, it adapts to your way of working, showing the commands you use most frequently in the suggestion list.

SYNONYM SUGGESTIONS

The new command line in AutoCAD 2014 has a built in **Synonym list**. When you enter a word at the command line, AutoCAD returns a command if it can match it in the synonym list.

For example, if you enter **BREAKUP** at the command line, AutoCAD will return the command **EXPLODE**. Or if you want to type a paragraph of text and enter the word **PARAGRAPH** at the command line, AutoCAD will return the command **MTEXT**.

You can add your own synonym's to the list which can be accessed by selecting:

Ribbon = Manage Tab / Customization Panel / Edit Aliases ▼ / Edit Synonym List

INTERNET SEARCH

AutoCAD 2014 now allows you to search for more information on a command that is displayed in the suggestion list. If you move the mouse cursor over a command in the list, it will display a **"Search on Internet"** icon and a **"Search in Help"** icon.

You can click on either of these two options to get extended help on the command you entered. So for example, if you type in **ARRAY** at the command line and choose the **Search on Internet** option, your current internet browser will open and show internet suggestions for **AutoCAD ARRAY**.

Continued on the next page...

Whichever command you choose to search for help on the internet, the word **AutoCAD** will always precede it. An example of the search and help icons is shown below.

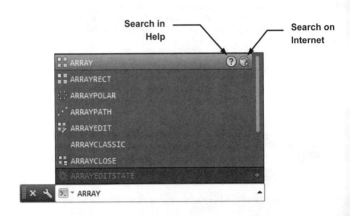

Search in
Help

Search on
Internet

CONTENT

The new Command Line in AutoCAD 2014 also allows you to quickly access layers, blocks, hatch patterns/gradients, text styles, dimension styles and visual styles. For example, if you have a drawing open that has block definitions with the name **DOOR**, and you enter **DOOR** at the command line, the suggestion list will display all the blocks with that name in it, so you could then insert that block directly from the command line.

The example below shows the command line suggestion list with block definitions of **DOOR**, you would simply click on the block you needed, and then insert it into your drawing.

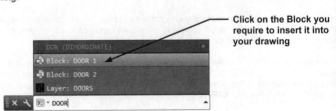

Click on the Block you
require to insert it into
your drawing

CATEGORIES

The command line suggestion list is made easier to navigate by organizing commands and system variables into **categories**. To see the results you can expand the category by clicking on the **+** sign, or you can press the **Tab** key to cycle through each category.

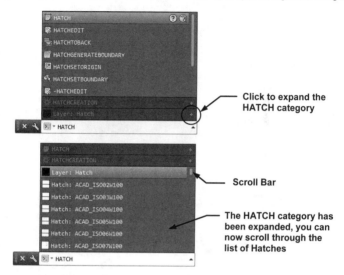

Click to expand the HATCH category

Scroll Bar

The HATCH category has been expanded, you can now scroll through the list of Hatches

INPUT SETTINGS

You can choose to turn on or off any of the new command line features by right-clicking on the command line and selecting **Input Settings**, you can choose between **AutoComplete, AutoCorrect, Search System Variables, Search Content,** and **Mid-string Search**.

Continued on the next page...

In addition to the **Input Settings**, you can further refine the settings by right-clicking on the command line and selecting **"Input Search Options"**. This will open the **Input Search Options** dialog box where you can change settings like the amount of times you can mistype a command before it gets entered into the **AutoCorrect List**.

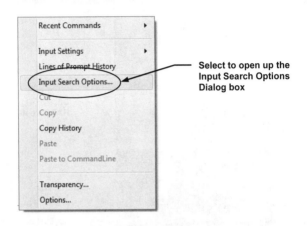

Select to open up the
Input Search Options
Dialog box

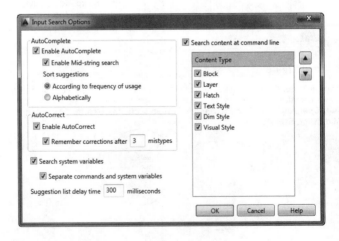

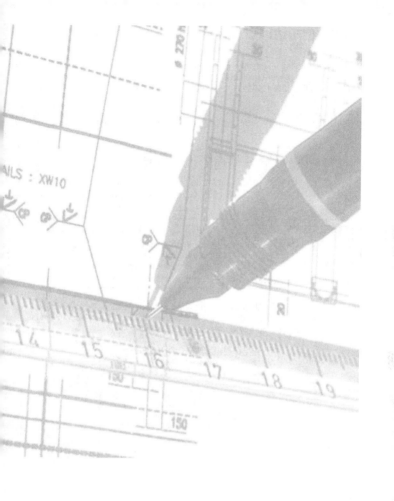

Section 9
Plotting

BACKGROUND PLOTTING

Background Plotting allows you to continue to work while your drawing is plotting. This is a valuable time saver because some drawings take a long time to plot. Or maybe you have multiple drawings to plot and you do not want to tie up your computer.

If you wish to view information about the status of the plot, click on the plotter icon located in the lower right corner of the AutoCAD window.

When the plot is complete, a notification bubble will appear.

If you do not want this bubble to appear you may turn it off. Right click on the plotter icon and select **"Enable Balloon Notification"**. This will remove the check mark and turn the notification off. You may turn it back on using the same process.

When you click on the **"Click to view plot and publish details..."** , on the bubble, you will get a report like the one shown below. The report will list details about all of the drawings plotted in the current drawing session.

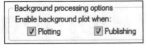

You may turn Background Plotting **ON** or **OFF**. The default setting is **OFF**.
To turn it **ON** or **OFF**:
Select: **Tools / Options / Plot and Publish Tab**.

PLOTTING FROM MODEL SPACE

1. **Important:** Open the drawing you want to plot, if it is not already open.
2. Select: **Zoom / All** to center the drawing within the plot area.
3. Select the **Plot** command using one of the following;

Quick Access Toolbar =
or
Ribbon = Output Tab / Plot Panel /
or
Keyboard = Plot <enter>

*The **Plot –Model** dialog box will appear.*

> **Note:** Another quick method of selecting the **Plot** command is by Right Clicking on the **Model Tab** and then select **Plot** from the menu.
>
>

4. Select the **"More Options"** button to expand the dialog box.

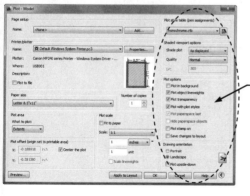

The dialog box expands to show more options.

Continued on the next page...

PLOTTING FROM MODEL SPACE....continued

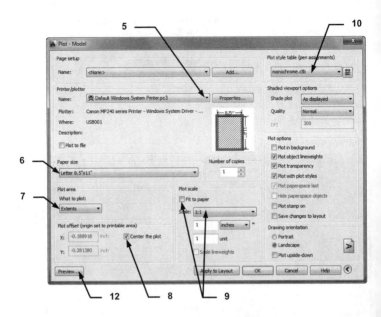

5. Select your printer from the drop down list or *"Default windows system printer.pc3"*
   Note: If your printer is not shown in the list you should configure your printer within AutoCAD. This is not difficult. Refer to Section 5 - **Add a Printer / Plotter**, for detailed instructions.

6. Select the Paper size such as; **Letter 8.5"x11"**

7. Select the Plot Area; **Extents**

8. Select the Plot Offset; **Center the Plot**

9. Uncheck the **"Fit to paper box"** and select Plot Scale **1:1**

10. Select the Plot Style table; **Monochrome.ctb** (for all black)
    Acad.ctb (for color)

Continued on the next page...

PLOTTING FROM MODEL SPACE....continued

11. If the following dialog box appears, select **Yes**.

12. Select the **Preview** button.

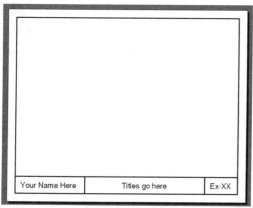

| Your Name Here | Titles go here | Ex-XX |

Does your display appear as shown above?
If **yes**, press **<enter>** and proceed to 13.
If not, recheck 1 through 11.

You have just created a **Page Setup**. All of the settings you have selected can now be saved. You will be able to recall these settings for future plots using this page setup. To save the **Page Setup** you need to **ADD** it to the model tab within this drawing.

13. Select the **ADD** button.

Continued on the next page...

PLOTTING FROM MODEL SPACE....continued

14. Type the new Page Setup name such as; **Model-Mono**.

(This name identifies that you will use it when plotting the Model Tab in Monochrome.)

15. Select the **OK** button.

14 → New page setup name: Model-Mono

15 → OK   Cancel   Help

The new Page Setup name appears

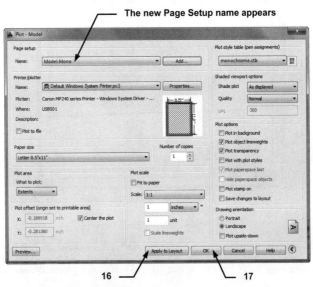

16. Select the **Apply to Layout** button.

17. Select the **OK** button to send the drawing to the printer or select <u>Cancel</u> if you do not want to print the drawing at this time. The Page Setup will still be saved.

18. Save the entire drawing again. The Page Setup will be saved to the **Model Tab** within the drawing and available to select in the future. You will not have to select all the individual settings again unless you choose to change them.

PLOTTING FROM PAPER SPACE

1. Open the **Drawing** you wish to plot.

2. Select the **Layout Tab** you wish to plot.

3. Select the **Plot** command using one of the following:

Quick Access Toolbar =
or
Ribbon = Output Tab / Plot Panel /
or
Application Menu = Print / Plot
or
Keyboard = Plot <enter>

The Plot dialog box shown below should appear.
Select the "More Options" ⊙ *button in the lower right corner*
if your dialog box does not appear the same as shown below.

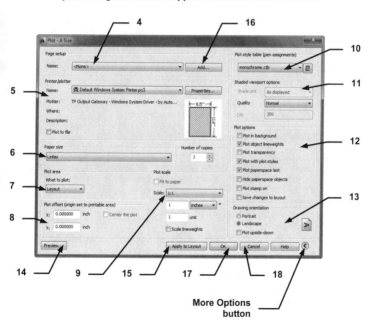

**More Options
button**

Continued on the next page...

PLOTTING FROM PAPER SPACE....continued

4. **Page Setup name:**
   After you have selected the desired settings you will save the new page setup and it will appear here. If you have previously created a page setup you may select it from the drop down list and all of the settings will change to reflect the previously saved page setup settings.

5. **Printer / Plotter:**
   Select the Printer that you wish to use. All previously configured devices will be listed here. (If your printer / plotter is not listed, refer to Section 5 - **Add a Printer / Plotter**.)

6. **Paper Size:**
   Select the paper size. The paper sizes shown in the drop down list are the available sizes for the printer that you selected. If the size you require is not listed, the printer you selected may not be able to handle that size. For example, a letter size printer can not handle a 24 X 18 size sheet. You must select a large format printer.

7. **Plot Area:**
   Select the area to plot. Layout is the default.

   Limits   plots the area inside the drawing limits.
             (This option is only available when plotting from model space.)

   Layout   plots the paper size.
             (Select this option when plotting from a Layout.)

   Extents   plots all objects in the drawing file even if out of view.
             (This option only available if you have a viewport.)

   Display   plots the drawing exactly as displayed on the screen.

   Window   plots objects inside a window. To specify the window, choose **Window** and specify the first and opposite (diagonal) corner of the area you choose to plot. (Similar to the **Zoom / Window** command.)

8. **Plot offset:**
   The plot can be moved away from the lower left plot limit corner by changing the **X** and / or **Y offset**.
   (If you have select Plot area "Display" or "Extents", select **"Center the plot."**)

9. **Scale:**
   Select a **scale** from the drop down list or enter a custom scale.

   Note: This scale is the Paper Space scale. The Model space scale will be adjusted within the viewport. If you are plotting from a **"LAYOUT"** tab, normally you will use **Plot Scale 1:1**

Continued on the next page...

PLOTTING FROM PAPER SPACE....continued

10. **Plot Style Table:**
    Select the Plot Style Table from the list. The Plot Styles determine if the plot is in color, Black ink or screened. You may also create your own.
    If you want to print in Black Ink only select **Monochrome.ctb**
    If you want to print in Color select **Acad.ctb**

11. **Shaded viewport options:**
    This area is used for printing shaded objects when working in the 3D environment.

12. **Plot options:**
    Plot background = specifies that the plot is processed in the background.

    Plot Object Lineweights = plots objects with assigned lineweights.

    Plot transparency = Plots any transparencies

    Plot with Plot Styles = plots using the selected Plot Style Table.

    Plot paperspace last = plots model space objects before plotting paperspace objects. Not available when plotting from model space.

    Hide Paperspace Objects = used for 3D only. Plots with hidden lines removed.

    Plot Stamp on = Allows you to print information around the perimeter of the border such as; drawing name, layout name, date/time, login name, device name, paper size and plot scale.

    Save Changes to Layout = Select this box if you want to save all of these settings to the current Layout tab.

13. **Drawing Orientation:**
    Portrait = the short edge of the paper represents the top of the page.

    Landscape = the long edge of the paper represents the top of the page.

    Plot Upside-down = Plots the drawing upside down.

14. Select the **Preview** button.
    The preview displays the drawing as it will plot on the sheet of paper.

*(Note: If you cannot see through to Model space, you have not cut your **Viewport** yet.)*

If the drawing appears as you would like it, press the **Esc** key and continue.

If the drawing does not look correct, press the **Esc** key and re-check your settings, then preview again.

(Note: If you have any of the layers set to "No Plot" they will not appear in the preview display. The preview display only displays what will be printed.)

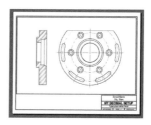

Continued on the next page...

PLOTTING FROM PAPER SPACE....continued

15. **Apply to Layout button:**

    This applies all of the settings to the layout tab. Whenever you select this layout tab the settings will already be set.

16. **Save the Page Setup:**

    At this point you have the option of saving these settings as another page setup for future use on other layout tabs. If you wish to save this setup, select the **ADD** button, type a name and select **OK**.

17. If your computer <u>**is**</u> connected to the plotter / printer selected, select the **OK** button to plot, then proceed to 19.

18. If your computer <u>**is not**</u> connected to the plotter / printer selected, select the **Cancel** button to close the Plot dialog box and proceed to 19.

    Note: Selecting Cancel <u>will cancel</u> your selected setting if you did not save the page setup as specified in 16 above.

19. **Save the drawing:**

    This will guarantee that the **Page Setup** you just created will be saved to this file for future use.

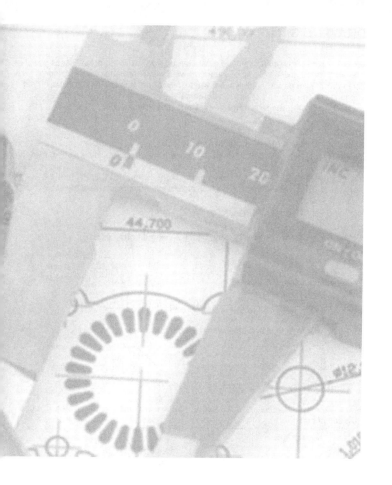

Section 10
Settings

DRAWING SETUP

When drawing with a computer, you must **"set up your drawing area"** just as you would on your drawing board if you were drawing with pencil and paper. You must decide what **size paper** you will need, what **Units of measurement** you will use (feet and inches or decimals, etc) and how **precise** you need to be. In CAD these decisions are called **"Setting the Drawing Limits, Units and Precision"**.

DRAWING LIMITS
Consider the drawing limits as the **size of the paper** you will be drawing on.
You will first be asked to define where the **lower left corner** should be placed, then the **upper right corner**, similar to drawing a Rectangle. An 11 x 8.5 piece of paper would have a **lower left corner** of **0,0** and an **upper right corner** of **11, 8.5**. *(11 is the horizontal measurement X-axis, and 8.5 is the vertical measurement Y-axis.)*

HOW TO SET THE DRAWING LIMITS

1. Select the **DRAWING LIMITS** command by typing: **Limits <enter>**.

2. The following prompt will appear:

   Command: '_limits
   Reset Model space limits:
   LIMITS Specify lower left corner or [**ON/OFF**] <0.0000,0.0000>:

   — Displays the current lower left corner coordinates <u>before</u> the change

3. Type the **X,Y** coordinates **0,0 <enter>** for the new **lower left corner** location of your piece of paper.

4. The following prompt will appear:

   LIMITS Specify upper right corner <12.0000,9.0000>:

   — Displays the current upper right corner coordinates <u>before</u> the change

5. Type the **X,Y** coordinates **11,8.5 <enter>** for the new **upper right corner** of your piece of paper.

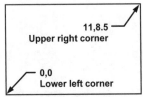

11,8.5
Upper right corner

0,0
Lower left corner

Note: visually the screen has not changed. Do the next step and it will.

This next step is important:

6. Type **Z <enter> A <enter>** to make the screen display the new drawing limits.
   (This is the shortcut for **Zoom / All**.)

DRAWING SETUP....continued

Grids within Limits:

If you have your **Grid behavior** setting ☐ **Display grid beyond Limits** is turned **Off** (no check mark) the grids will only be displayed within the Limits that you set.

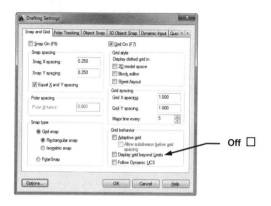

Off ☐

Grids displayed within Limits only

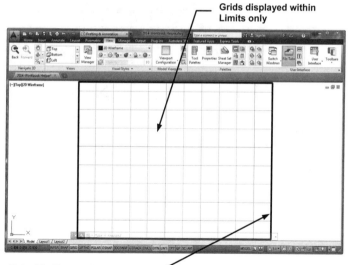

Note: The Rectangular border is to indicate the outline of the limits. It will not appear normally.

DRAWING SETUP....continued

Grids beyond Limits:

If you have your **Grid behavior** setting ☑ **Display grid beyond Limits** is turned **On** (check mark) the grids will be displayed beyond the Limits that you set.

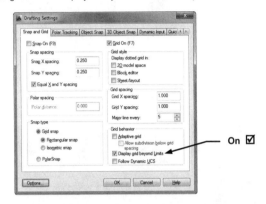

On ☑

Grids displayed beyond Limits.

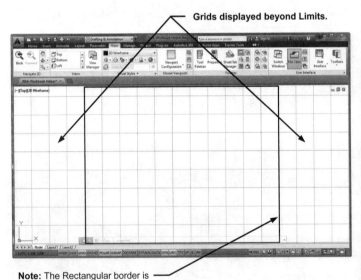

Note: The Rectangular border is to indicate the outline of the limits. It will not appear normally.

DRAWING SETUP....continued

UNITS AND PRECISION:

You now need to select what **unit of measurement** with which you want to work.
Such as: Decimal (0.000) or Architectural (0'-0").
Next you should select how precise you want the measurements. For example, do you
want the measurement limited to a 3 place decimal or the nearest 1/8".

HOW TO SET THE UNITS AND PRECISION.

1. Select the **UNITS** command using one of the following:

   Application Menu = Drawing Utilities / Units
   or
   Keyboard = Units <enter>

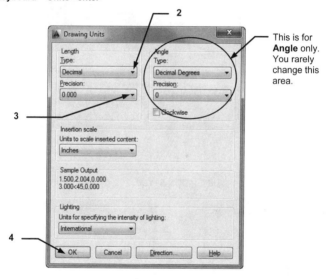

This is for
Angle only.
You rarely
change this
area.

2. **Length Type:** Select the down arrow and select: **Decimals** or **Architectural**.

3. **Length Precision:** Select the down arrow and select the appropriate **Precision**
   associated with the "type".
   Examples: 0.000 for Decimals or 1/16" for Architectural.

4. Select the **OK** button to save your selections.

ANNOTATIVE PROPERTY

The **"Annotative" property** automates the process of scaling text, dimensions, hatch, tolerances, leaders and symbols. The height of these objects will automatically be determined by the annotation scale setting.

You can add an **"Annotative" property** to dimension styles and text styles.

TEXT STYLE

This symbol △ indicates Annotative

Select the Annotative option here

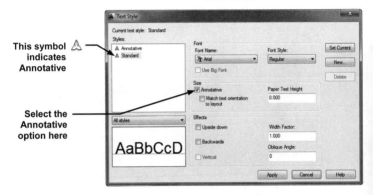

DIMENSION STYLE

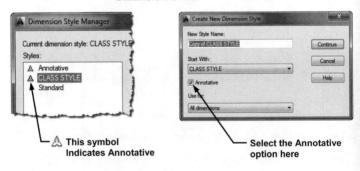

△ This symbol Indicates Annotative

Select the Annotative option here

ANNOTATIVE OBJECTS

You can add the **"Annotative" property** to Text or Dimension Styles simply by placing a check mark in the Annotative box.

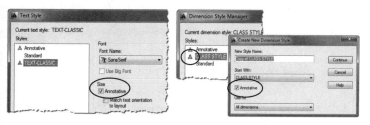

The Text style and Dimension Style shown above are now **Annotative**.
Notice the Annotative symbol beside the Style Name. (No symbol by "Standard".)

"Annotative" Objects are scaled automatically to match the scale of the viewport. For example, if you want the text, inside the viewport, to print .200 in height and the viewport scale is 1:2 AutoCAD will automatically scale the text height to .400. The text height needs to be scaled by a factor of 2 to compensate for the model space contents appearing smaller.

The easiest way to understand how **"Annotative" property** works is to do it.

So try this example:

1. Start a **New** drawing file.
2. Select the **Model** tab.
3. Draw a Rectangle; **3.00"** Long **X 1.50"** Wide. *(Use a Layer named Object Line.)*

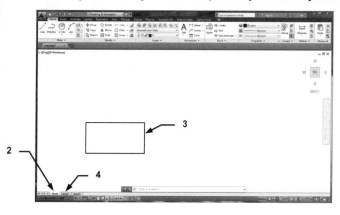

4. Select a **Layout Tab**.

Continued on the next page...

ANNOTATIVE OBJECTS....continued

5. **Erase** any existing **viewport frames**.

6. **Cut 2 new viewports** as shown below. *(Use a Layer named Viewport.)*

7. Activate the **left viewport** (double click inside the viewport frame) and do the following:

   A. Adjust the **Scale** of the viewport to **1:1**
   B: **PAN** to place the rectangle in the **center** of the viewport.
   C. **Lock** the viewport.

8. Activate the **right viewport** (click inside the viewport frame) and do the following:

   A. Adjust the **Scale** of the viewport to **1:4**
   B: **PAN** to place the rectangle in the **center** of the viewport.
   C. **Lock** the viewport.

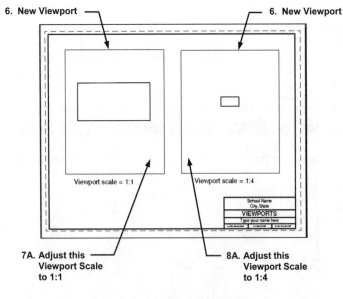

Your screen should appear approximately like this.

Continued on the next page...

ANNOTATIVE OBJECTS....continued

9. Activate the **left hand** viewport. (Double click inside the left hand viewport)

10. Change current layer to **Text** and add **.200"** height text, **Scale 1:1** as shown using text style "Text-Classic". (Note: Text style "Text-Classic" is Annotative)

11. Change current layer to **Dimension** and add the dimension shown using dimension style "Dim-Decimal". (Note: Dimension style "Dim-Decimal" is Annotative)

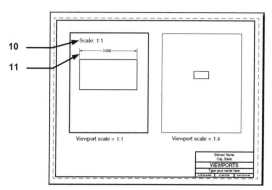

12. Activate the **right hand** viewport. (Click inside the right hand viewport)

13. Change current layer to **Text** and add **.200"** height text, **Scale 1:4** as shown using text style "Text-Classic". (Note: Text style "Text-Classic" is annotative)

14. Change current layer to **Dimension** and add the dimension shown using dimension style "Dim-Decimal". (Note: Dimension style "Dim-Decimal" is annotative)

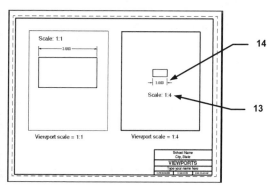

Continued on the next page...

ANNOTATIVE OBJECTS....continued

Notice that the text and dimensions appear the <u>same size in both viewports</u>.

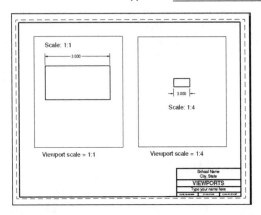

15. Now select the **Model** Tab.

Notice there are 2 sets of text and dimensions.

One set has the **Annotative Scale** of **1:1** and will be visible only in a **1:1 Viewport**.
One set has the **Annotative Scale** of **1:4** and will be visible only in a **1:4 Viewport**.
But you see both sets when you select the model tab.

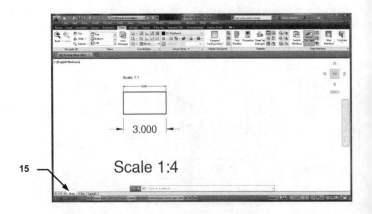

Continued on the next page...

ANNOTATIVE OBJECTS....continued

16. Place your cursor on any of the text or dimensions. An **"Annotative symbol"** will appear. This indicates this object is annotative and it has only one annotative scale.

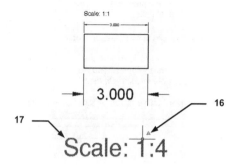

17. Click once on the **Scale: 1:4** text. The Quick properties box should appear. (**QP** button on the status bar must **ON**.)

Notice the text height is listed twice.

Paper text height = .200   This is the height that you selected when placing the text. When the drawing is printed the text will print **.200**

Model text height = .800   This is the desired height of the text (.200) factored by the viewport scale (1:4). The viewport scale is a factor of 4. (4 X .200 = .800)

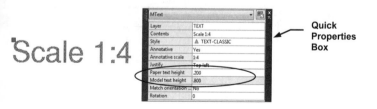

Quick
Properties
Box

Summary:

If an object is **Annotative**, AutoCAD automatically adjusts the scale of the object to the viewport scale. The most commonly used Annotative Objects are: Dimensions, Text, Hatch and Multileaders. Refer to the Help menu for more

ASSIGNING MULTIPLE ANNOTATIVE SCALES

The previous pages showed you how annotative text and dimensions are automatically scaled to the viewport scale. But in order to have an annotative object appear in both viewports you placed 2 sets of text and dimensions. Now you will learn how to easily assign multiple annotative scales to a single text string or dimension so you need not duplicate them each time you create a new viewport. You will just assign an additional annotative scale to the annotative object.

Again, the easiest way to understand this process is to do it. The following is a step by step example.

1. Use the example from the previous pages. *If you did not complete the example from the previous pages, it is advisable to complete them now.*

2. Make the **right hand viewport active**. (Double click inside the viewport)

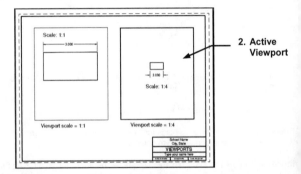

2. Active Viewport

3. Erase the <u>text</u> and the <u>dimension</u> in the **right hand** viewport only.

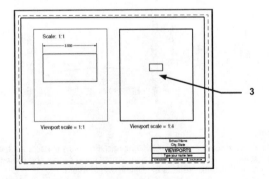

3

Continued on the next page...

ASSIGNING MULTIPLE ANNOTATIVE SCALES....continued

4. Display all annotative objects in all viewports as follows:

   A. Select the **Annotation Visibility** button located in the lower right corner of the drawing status bar.

4A. Annotation Visibility button.

Light bulb should Be Yellow.

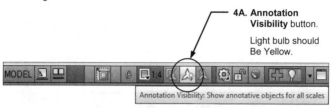

Annotation Visibility: Show annotative objects for all scales

 ON (**Yellow** light bulb) Displays **all** Annotation objects in **all** Viewports. (Example below)

 OFF (**Blue** light bulb) Displays **only** Annotation objects that have an Annotative Scale that matches the Viewport Scale.

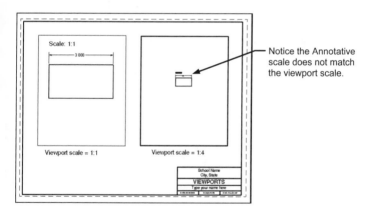

Notice the Annotative scale does not match the viewport scale.

The dimensions and text are now displayed in both viewports. But the annotative scale of the dimensions and text in the right hand viewport do not match the scale of the viewport. (Notice they are smaller) **The scale of annotative objects must match the scale of the viewport**.

Follow the steps on the next page to assign multiple annotative scales to an annotative object.

Continued on the next page...

ASSIGNING MULTIPLE ANNOTATIVE SCALES....continued

5. Place your cursor near the **dimension** in the right hand viewport. Notice the **single Annotation symbol**. This single symbol indicates the annotative dimension has only one annotative scale assigned to it.

6. Select the **Annotate** Tab.

7. Select **only** the dimension in the **right-hand viewport**.

8. Select the **Add Current Scale** tool on the **Annotation Scaling** panel.

The **Annotative dimension** should have increased in size as shown below.

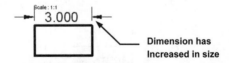

Dimension has
Increased in size

9. Turn **OFF** the Annotation Visibility. (Click on the button. The light should turn blue)

9. **Annotation Visibility**
button should turn Blue

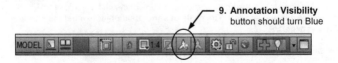

Continued on the next page...

ASSIGNING MULTIPLE ANNOTATIVE SCALES....continued

Notice the text in the right hand viewport is no longer visible. When the **Annotative Visibility** is **OFF** only the annotative objects that match the viewport scale will remain visible. The dimension is the only annotative object that has the 1:4 annotative scale assigned to it.

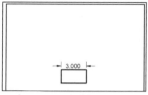

Annotative Visibility OFF

10. Place your cursor near the dimension. Notice **2 Annotation symbols** appear now. This indicates 2 annotation scales have been assigned to the annotative object.

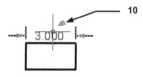

11. Click on the dimension to display the grips and drag the dimension away from the rectangle approximately as shown below. (The dimension was too close to the rectangle) **Notice the dimension in the left hand viewport did not move**. They can be moved individually.

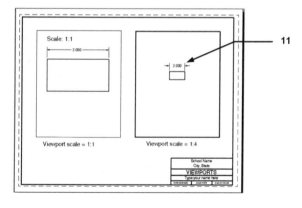

HOW TO REMOVE AN ANNOTATIVE SCALE

If you have an annotative object such as a dimension, that you would like to remove from a viewport, you must remove the annotative scale that matches the viewport scale. **Do not delete** the dimension because it will also be deleted from all of the other viewports. This sounds complicated but is very easy to accomplish.

Problem: I would like to remove **Dimension A** from the right hand viewport but I do not want **Dimension B** in the lower viewport to disappear.

Solution: I must remove the **1:4** Annotation scale from **Dimension A**.

(Refer to Step by step instructions below.)

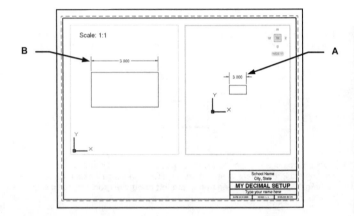

Step 1:

1. Select the **Annotate** tab on the Ribbon.

2. Select **Dimension A** shown above. (You must be inside the viewport)

3. Select the **Add / Delete Scales** tool located on the **Annotation Scaling** panel.

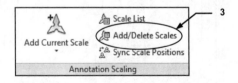

Continued on the next page...

HOW TO REMOVE AN ANNOTATIVE SCALE....continued

4. Select the Annotation Scale to remove. (**1:4**)

5. Select the **Delete** button.

   (Remember, you are deleting the annotative scale from the dimension. You are not deleting the dimension. The dimension still exists but it will not have an annotative scale of 1:4 assigned to it. As a result it will not be visible within any viewport that has been scaled to 1:4)

6. Select the **OK** button.

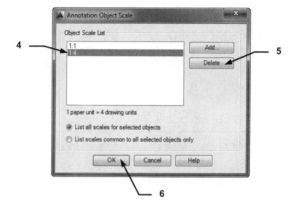

Note: Dimension A has been removed and **Dimension B** remains.

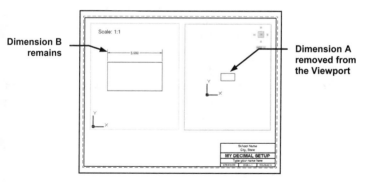

Dimension B remains

Dimension A removed from the Viewport

ANNOTATIVE HATCH

Hatch may be **Annotative** also. You may select the Annotative setting as you create the Hatch set or you may add the Annotative setting to an existing Hatch set.

How to select the Annotative setting as you create the Hatch set.

1. Select the **Hatch** command and select the desire settings including **Annotative**. The Hatch set created will be annotative.

How to change an existing Non-Annotative Hatch set to Annotative.

1. Click on the **existing** Hatch set.

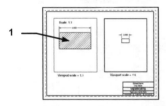

The **Hatch Editor** Ribbon Tab with appear.

2. Select the **Annotative** option.

3. Select **Close Hatch Editor**.

The Non-Annotative Hatch is now Annotative.

ANNOTATIVE HATCH....continued

How to assign multiple Annotative scales to a Hatch set.

1. Draw the hatch in one of the viewports using **Annotative** hatch.
   (Refer to previous page)

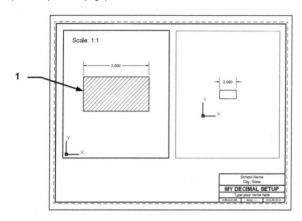

2. Turn **ON Annotation Visibility** (Yellow) (Refer to page 10-13)

3. Select the **Hatch Set** to change.

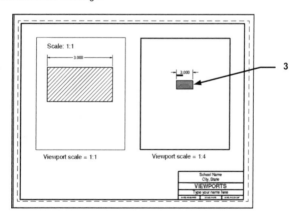

Continued on the next page...

ANNOTATIVE HATCH....continued

4. Select the **Annotate Tab**.

5. Select the **"Add Current Scale"** tool.

6. Turn **OFF Annotation Visibility**. (Blue) (Refer to page 10-13)

Now the appearance of the hatch sets should be identical in both viewports.

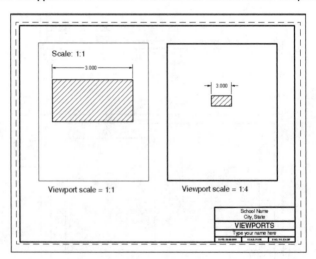

Section 11
Text

CREATING NEW TEXT STYLES

AutoCAD provides you with two preset Text Styles named "Standard" and "Annotative".
You may want to create a new text style with a different font and effects.
The following illustrates how to create a new text style.

1. Select the **TEXT STYLE** command using one of the following:

   Ribbon = Annotate Tab / Text Panel / ↘
   or
   Keyboard = ST <enter>

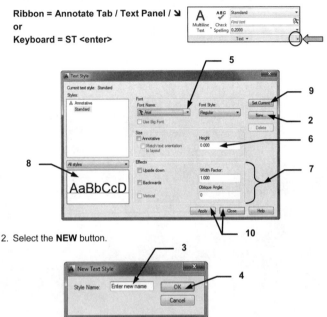

2. Select the **NEW** button.

3. Type the new style name in **STYLE NAME** box.
   Styles names can have a maximum of 31 characters, including letters, numbers,
   dashes, underlines and dollar signs. You can use Upper or Lower case.

4. Select the **OK** button.

5. Select the **FONT**.

6. Enter the value of the **Height**.

Note: If the value is **0**, AutoCAD will always prompt you for a height.
If you enter a number the new text style will have a fixed height and AutoCAD will not
prompt you for the height.

Continued on the next page...

CREATING NEW TEXT STYLES....continued

7. Assign **EFFECTS**.

Effects	
☐ Upside down	Width Factor: 1.000
☐ Backwards	Oblique Angle:
☐ Vertical	0

UPSIDE-DOWN:
Each letter will be created upside-down in the order in which it was typed.
(**Note:** this is different from rotating text **180** degrees.)

BACKWARDS:
The letters will be created backwards as typed.

VERTICAL:
Each letter will be inserted directly under the other. Only "**.shx**" fonts can be used.
VERTICAL text will not display in the **PREVIEW** box.

OBLIQUE ANGLE:
Creates letter with a slant, like italic. An angle of **0°** creates a vertical letter. A positive angle will slant the letter forward. A negative angle will slant the letter backward.

WIDTH FACTOR:
This effect compresses or extends the width of each character.
A value less than **1** compresses each character.
A value greater than **1** extends each character.

8. The **PREVIEW** box displays the text with the selected settings

9. Select the **Set Current** button.

10. Select the **Apply** or **Close** button.

HOW TO SELECT A TEXT STYLE

After you have created Text Styles you will want to use them. You must select the Text Style before you use it.

Below are the methods of selection when using **Single Line** or **Multiline text**.

SINGLE LINE TEXT:

Select the style before selecting the Single Line Text command.

1. Select the **Annotate** Tab.
2. Using the Text panel, select the **style down arrow** ▾
3. Select the **Text Style**.

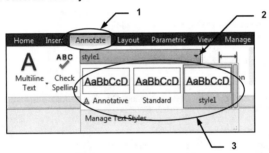

MULTILINE TEXT:

1. Select **Multiline Text** and place the **first corner** and **opposite corner**.
2. Find the **Style** panel and scroll through the text styles available using the **up** and **down** arrows.

Select the **Text Style**
Within the **Text Editor**.

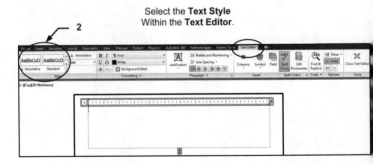

DELETE A TEXT STYLE

1. Select the **TEXT STYLE** command using one of the following:

Ribbon = Annotate Tab / Text Panel / ↘
or
Keyboard = ST <enter>

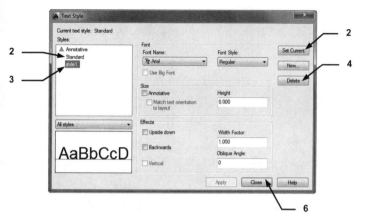

2. First, select a **Text Style** that you <u>do not want to Delete</u> and select the **Set Current** button. (Note: You can't Delete a Text Style that is in use.)

3. Select the **Text Style** that you want to Delete.

4. Select the **Delete** button.

5. A warning appears, select **OK** or **Cancel**.

6. Select the **Close** button.

Note: Also refer to the **PURGE** command. The **Purge** command will remove any unused **text styles**, **dimension styles**, **layers** and **linetypes**.

CHANGE EFFECTS OF A TEXT STYLE

1. Select the **TEXT STYLE** command using one of the following:

Ribbon = Annotate Tab / Text Panel / ↘
or
Keyboard = ST <enter>

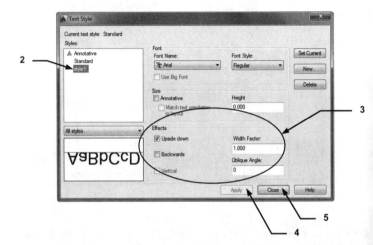

2. Select the **Text Style** you wish to change.

3. Make the changes in the **EFFECTS** boxes.

   Note about Vertical:
   Only "**.shx**" fonts can be vertical.
   Vertical text will not display in the **PREVIEW** box.

4. Select the **Apply** button. (*Apply will stay gray if you did not change a setting*.)

5. Select the **Close** button.

MULTILINE TEXT

MULTILINE TEXT command allows you to easily add a sentence, paragraph or tables. The Mtext editor has most of the text editing features of a word processing program. You can underline, bold, italic, add tabs for indenting, change the font, line spacing, and adjust the length and width of the paragraph.

When using Mtext you must first define a text boundary box. The text boundary box is defined by entering where you wish to start the text (first corner) and approximately where you want to end the text (opposite corner). It is very similar to drawing a rectangle. The paragraph text is considered one object rather than several individual sentences.

USING MULTILINE TEXT

1. Select the **MULTILINE TEXT** command using one of the following:

Ribbon = Annotate Tab / Text Panel /
or
Keyboard = MT <enter>

The command line will list the current style, text height and annotative setting.

Mtext Current text style: "STANDARD"   Text height: .200   Annotative: No

The cursor will then appear as crosshairs with the letters "abc" attached. These letters indicate how the text will appear using the current font and text height.

2. Specify first corner: *Place the cursor at the upper left corner of the area where you want to start the new text boundary box and press the left mouse button. (P1)*

3. Specify opposite corner or [Height / Justify / Line Spacing / Rotation / Style / Width / Columns]: *Move the cursor to the right and down (P2) and press left mouse button.*

Continued on the next page...

MULTILINE TEXT....continued

The Text Editor Ribbon will appear.

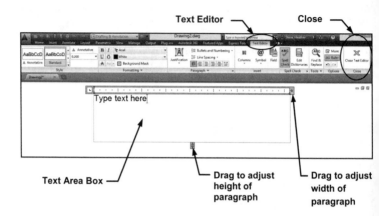

Text Editor — Close —

Text Area Box — Drag to adjust height of paragraph — Drag to adjust width of paragraph

The **Text Editor** allows you to select the Text Style, Font, Height etc. You can add features such as bold, italics, underline and color.

The **Text Area box** allows you to enter the text, add tabs, indent, adjust left hand margins and change the width and height of the paragraph.

4. After you have entered the text in the Text Area box, select the **Close Text Editor** tool.

HOW TO CHANGE THE "abc", ON THE CROSSHAIRS, TO OTHER LETTERS.

You can personalize the letters that appear attached to the crosshairs using the **MTJIGSTRING** system variable. (10 characters max) The letters will simulate the appearance of the font and height selected but will disappear after you place the lower right corner (**P2**).

1. Type **MTJIGSTRING <enter>** on the command line or dynamic input box.
2. Type the new letters followed by **<enter>**.

The letters will be saved to the computer, not the drawing. They will appear anytime you use **Mtext** and will remain until you change them again.

TABS, INDENTS AND SPELLING CHECKER

TABS

Setting and removing Tabs is very easy.

The increments are determined by the text height. For example: If the text height is 1" you may quickly place a tab at any 1" mark on the ruler.

Set or change the stop positions at anytime, using one of the following methods.

Place the cursor on the "Ruler" where you want the tab and left click. A little dark **"L"** will appear. The tab is set. If you would like to remove a tab, just click and drag it off the ruler and it will disappear.

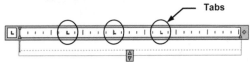

INDENTS

Sliders on the ruler show indention relative to the left side of the text boundary box. The top slider indents the first line of the paragraph, and the bottom slider indents the other lines of the paragraph.

You may change their positions at anytime, using one of the following methods.

Place the cursor on the **"Slider"** and click and drag it to the new location.

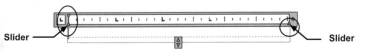

SPELLING CHECKER

If you have Spell check **ON** you will be alerted as you enter text with a **Red Line** under the misspelled word. Right click on the word and AutoCAD will give you some choices.

1. Select the text you wish to Spell check. (Click once on sentence)

2. Select **Annotate tab / Text panel /**

   The **Check Spelling** dialog
   box will appear.

3. Select **Start**.
   If AutoCAD finds any words misspelled it will suggest a change.
   You may select **Change** or **Ignore**.
   When finished a message will appear stating **"Spelling Check Complete"**.

COLUMNS

STATIC COLUMNS

1. Right click in the **Text Box Area** and select **Columns**.

2. Select **Column Settings…**

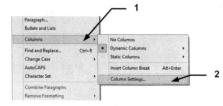

The **Column Settings** Dialog box appears.

3. Select **Static Columns**.

4. Select:
   A. Column Number
   B. Height
   C. Width
   D. Gutter

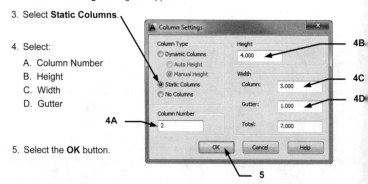

5. Select the **OK** button.

6. The Text Area should appear as shown below with 2 columns divided with a gutter.

7. Start typing in the left hand box. When you fill the left hand box the text will start to spill over into the right hand box.

You may also adjust and make changes to the width and height using the drag tools.

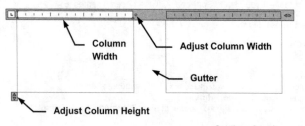

Column Width

Adjust Column Width

Gutter

Adjust Column Height

Continued on the next page...

COLUMNS....continued

DYNAMIC COLUMNS

1. Right click in the **Text Box Area** and select **Columns**.

2. Select **Column Settings...**

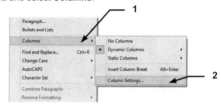

The **Column Settings** Dialog box appears.

3. Select **Dynamic Columns** and **Auto Height**

4. Select:

   A. Height
   B. Width
   C. Gutter

5. Select the **OK** button.

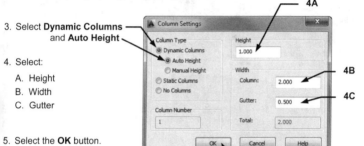

6. The Text Area will first appear with one column with the width and height you set.

7. When the text fills the first column another column will appear.

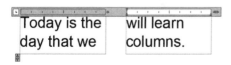

When the second column fills another column will appear.
You may also adjust and make changes to width and height using the drag tools.

PARAGRAPHS AND LINE SPACING

PARAGRAPH and LINE SPACING

You may set the **Tabs**, **Indent** and **Line Spacing** for individual paragraphs.

1. Right click in the **Text Box Area** and select **Paragraph**.

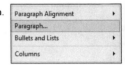

The **Paragraph** dialog box will appear.

You may add or remove tabs here also.

To Add:
1. Enter the **Spacing**.
2. Select **Add** button.

To Remove:
1. Select from the list.
2. Select **Remove** button.

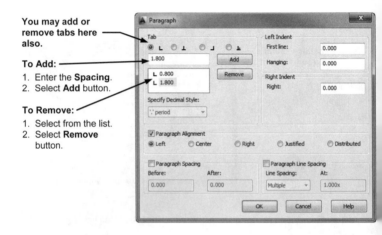

EDITING MULTILINE TEXT

MULTILINE TEXT

Multiline Text is as easy to edit as it is to input originally. You may use any of the text options shown on the **Text Editor** tab.

1. Double click on the Multiline text you want to edit.

2. Highlight the text, that you want to change, using click and drag.

3. Make the changes then select the **Close Text Editor** tool.

You may also edit many other Multiline Text features.

1. Double click on the Multiline Text you wish to edit.

2. Right click in the **Text Box Area**.

The **Menu** shown opposite will appear.

SINGLE LINE TEXT

SINGLE LINE TEXT allows you to draw one or more lines of text. The text is visible as you type. To place the text in the drawing, you may use the default **START POINT** (the lower left corner of the text), or use one of the many styles of justification described on the next page. Each line of text is treated as a separate object.

USING THE DEFAULT START POINT

1. Select the **SINGLE LINE TEXT** command using one of the following:

Ribbon = Annotate Tab / Text Panel /
or
Keyboard = DT <enter>

Command: _text
Current text style: "STANDARD" Text height: 0.200 Annotative: No
Specify start point of text or [Justify/Style]: *Place the cursor where the text should start and left click.*
Specify height <0.200>: *type the height of your text <enter>.*
Specify rotation angle of text <0>: *type the rotation angle then <enter>.*
Enter text: *type the text string; press enter at the end of the sentence.*
Enter text: *type the text string; press enter at the end of the sentence.*
Enter text: *type the next sentence or press <enter> to stop.*

USING JUSTIFICATION

If you need to be very specific, where your text is located, you must use the Justification option. For example if you want your text in the middle of a rectangular box, you would use the justification option **"Middle"**.

The following is an example of Middle justification.

1. Draw a Rectangle **6"** wide and **3"** high.
2. Draw a Diagonal **Line** from one corner to the other corner.

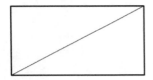

Continued on the next page...

SINGLE LINE TEXT....continued

3. Select the **SINGLE LINE TEXT** command.
   Command: _text
   Current text style: "STANDARD" Text height: 0.200
4. Specify start point of text or[Justify/Style]: *type J <enter>.*
5. Enter an option [Align/Fit/Center/Middle/Right/TL/ TC/TR/ML/MC/MR/BL/BC/BR]:
   type M <enter>.
6. Specify middle point of text: *snap to the midpoint of the diagonal line.*
7. Specify height <0.200>: *type 1 <enter>.*
8. Specify rotation angle of text <0>: *type 0 <enter>.*
9. Enter text: *type: HHHH <enter>.*
10. Enter text: *press <enter> to stop.*

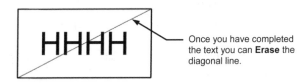

Once you have completed the text you can **Erase** the diagonal line.

Note:

An alternative method of placing the text in the center of the rectangle is by using the **"Mid Between 2 points"** Object Snap. By using this method you will not need to draw the diagonal line, you just simply snap to the 2 diagonal corners.

When you get to **Step 6** replace the method with the following alternative method:

6. Hold down the **Shift key** then **right click**.
   B. Select **"Mid Between 2 Points"** from the menu.
   C. Left click on **P1** for the first mid point.
   D. Left click on **P2** for the second mid point.
   E. Continue on to **Step 7.**

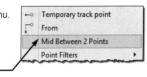

6B

P2

P1

SINGLE LINE TEXT....continued

<u>OTHER JUSTIFICATION OPTIONS</u>:

ALIGN
Aligns the line of text between two points specified.
The height is adjusted automatically.

FIT
Fits the text between two points specified.
The height is specified by you and does not change.

CENTER
This is a tricky one. Center is located at the bottom center of Upper Case letters.

MIDDLE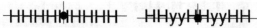
If only uppercase letters are used **MIDDLE** is located in the middle, horizontally and vertically. If both uppercase and lowercase letters are used **MIDDLE** is located in the middle, horizontally and vertically, but considers the lowercase letters as part of the height.

RIGHT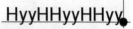
Bottom right of upper case text.

TL, TC, TR
Top left, Top center and Top right of upper and lower case text.

ML, MC, MR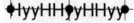
Middle left, Middle center and Middle right of upper case text.
(Notice the difference between "Middle" and "MC".)

BL, BC, BR
Bottom left, Bottom center and Bottom right of lower case text.
(Notice the different location for **BR** and **RIGHT** shown above.
BR considers the lower case letters with tails as part of the height.)

EDITING SINGLE LINE TEXT

EDITING SINGLE LINE TEXT

Editing **Single Line Text** is somewhat limited compared to Multiline Text. In the example below you will learn how to edit the text within a Single Line Text sentence.

1. Double click on the **Single Line Text** you want to **edit**. The text will highlight.

> ## This is Single Line Text.

2. Make the changes in place then press **<enter> <enter>**. (The second **<enter>** ends the command.)

YOU CAN ALSO USE THE PROPERTIES PALETTE TO EDIT SINGLE LINE TEXT.

1. Single click on the **Single Line Text** you want to **edit**.

2. Right click and then select **Properties** from the menu.

3. Make any changes necessary then press the **Esc** key to deselect the text.

4. **Close** the Properties Palette.

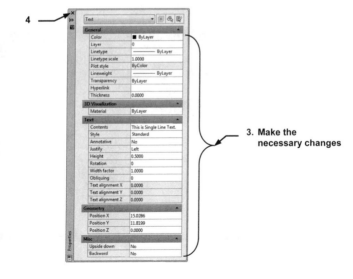

4

3. **Make the necessary changes**

SPECIAL TEXT CHARACTERS

Characters such as the **Degree symbol (°), Diameter symbol (Ø)** and the
Plus / Minus symbols (±) are created by typing **%%** and then the appropriate **"code"**
letter.

SYMBOL		CODE
Ø	Diameter	%%C
°	Degree	%%D
±	Plus / Minus	%%P

SINGLE LINE TEXT
When using "Single Line Text", type the code in the sentence. After you enter the code
the symbol will appear.

> **For example:**
> Entering **350%%D** will create: **350°**. The **"D"** is the **"code"** letter for degree.

MULTILINE TEXT
When using Multiline Text you may enter the code in the sentence or you may select a
symbol using the **Symbol Tool** located on the **Insert panel** of the **Text Editor** ribbon.

Symbol Tool

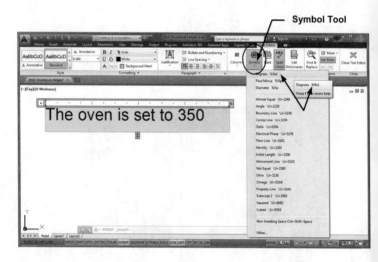

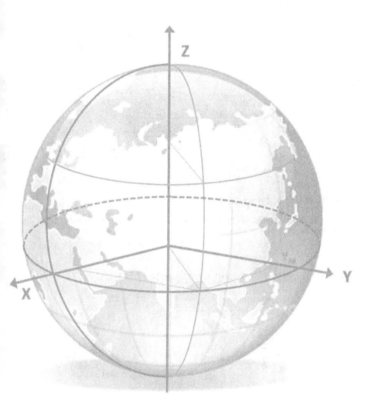

Section 12
UCS

DISPLAYING THE UCS ICON

The **UCS icon** is merely a drawing aid that displays the location of the Origin. It can move with the Origin or stay in the default location. You can even change its appearance.

SHOW UCS ICON AT ORIGIN

1. Right click on the **Origin icon**.
2. Select **UCS Icon Settings** from the shortcut menu.

3. **Show UCS Icon at Origin.**

    ☑ The UCS icon will follow the Origin as you move it. You will be able to see the Origin location at a glance.

    ☐ The UCS icon will not follow the Origin. It will stay in it's default location.

HOW TO CHANGE THE UCS ICON APPEARANCE

1. Right click on the **Origin icon**.
2. Select **UCS Icon Settings** from the short cut menu.

3. Select **Properties**.

When you select this option the dialog box shown opposite will appear. You may change the Style, Size and Color of the icon at any time. Changing the appearance is personal preference and will not affect the drawing or commands.

4. When complete select the **OK** button.

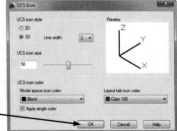

MOVING THE ORIGIN

The **ORIGIN** is where the **X**, **Y**, and **Z-axes** intersect. The Origin's (**0,0,0**) default location is in the lower left-hand corner of the drawing area. But you can move the Origin anywhere on the screen using the **UCS** command.
(The default location is designated as the **"World"** option or **WCS**. When it is moved it is **UCS**, User Coordinate System.)

You may move the Origin many times while creating a drawing. This is not difficult and will make it much easier to draw objects in specific locations.

Refer to the examples on the next page.

TO MOVE THE ORIGIN:

1. Right click on the **Origin icon**.

2. Select **Origin** from the shortcut menu.

3. Place the new Origin location by entering coordinates or pressing the left mouse button.

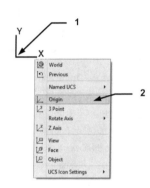

TO RETURN THE ORIGIN to the default "World" location (the lower left corner):

1. Right click on the **Origin icon**.

2. Select **World** from the shortcut menu.

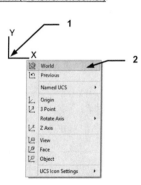

Why move the Origin?   Examples on the next page.

MOVING THE ORIGIN....continued

Why move the Origin?

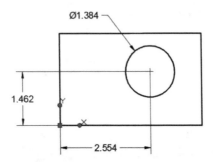

If you move the **Origin** to the lower left corner of the rectangle it will make it very easy to accurately place the center of the circle.

HOW TO PLACE THE CIRCLE ACCURATELY:

1. Select **"Origin"** from the shortcut menu as shown on the previous page.

2. Snap to the lower left corner of the rectangle using object snap **Endpoint**.

   (The UCS icon should now be displayed as shown above.)

3. Select the **Circle** command.

4. Enter the coordinates to the center of the circle: **2.554, 1.462 <enter>**

5. Enter the diameter: **1.384 <enter**

6. Select **"World"** from the shortcut menu to return the UCS icon to it's default location.
   (Refer to previous page.)

More examples of why you would move the Origin:

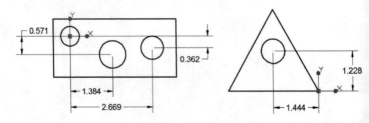

MOVING THE ORIGIN....continued

There is also another method which you can use to move the **UCS**, it is also much quicker. You can select and drag the **UCS** to a new location, simply by clicking on the Icon until you see three blue grips, if you left click on the square grip you can then drag the Icon and snap it to a new location. Press **Esc** to clear the grips.

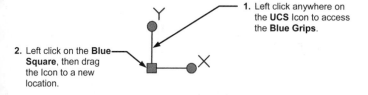

1. Left click anywhere on the **UCS** Icon to access the **Blue Grips**.

2. Left click on the **Blue Square**, then drag the Icon to a new location.

Another example of why you would move the Origin:

Consider the example below; the four outer circles are positioned from the bottom left-hand corner of the rectangle. The large inner circle is positioned from the top right-hand corner of the rectangle. And the four small circles are positioned from the large holes center-point.

To position all the circles accurately you could move the **UCS** to three different locations. The following page shows you how this is achieved.

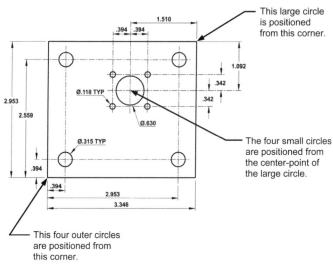

This large circle is positioned from this corner.

The four small circles are positioned from the center-point of the large circle.

This four outer circles are positioned from this corner.

Continued on the next page...

MOVING THE ORIGIN....continued

1. Select the **UCS** icon then left click on the **square grip**.

2. Move the **UCS** icon to position **P1**.

3. Create and position the four outer circles.

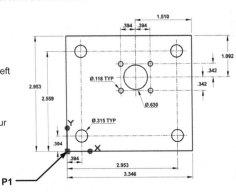

4. Select the **UCS** icon then left click on the **square grip**.

5. Move the **UCS** icon to position **P2**.

6. Create and position the large inner circle.

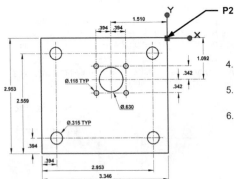

7. Select the **UCS** icon then left click on the **square grip**.

8. Move the **UCS** icon to position **P3**.

9. Create and position the four inner small circles.

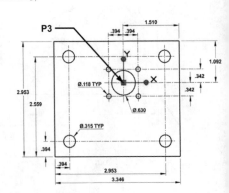

Section 13
Constraints

PARAMETRIC DRAWING

Parametric Drawing is a method of assigning a constraint to an object that is controlled by another object. For example, you could put a constraint on line #1 to always be parallel to line #2. Or you could put a constraint on diameter #1 to always be the same size as diameter #2. If you make a change to #1 AutoCAD automatically makes the change to #2 depending on which constraint has been assigned.

There are two general types of constraints:

<u>**GEOMETRIC:**</u> Controls the relationship of objects with respect to each other. The Geometric constraints that will be discussed in this workbook are coincident, collinear, concentric, fix, parallel, perpendicular, horizontal, vertical, tangent, symmetrical and equal.

Example:

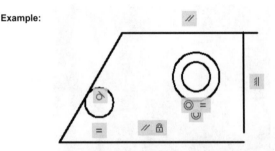

<u>**DIMENSIONAL:**</u> Dimensional constraints control the size and proportions of objects. They can constrain:
    A. Distances between objects or between points on objects.
    B. Angles between objects or between points on objects.
    C. Sizes of arcs and circles.

Example:

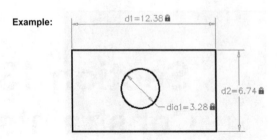

GEOMETRIC CONSTRAINTS

Overview of Geometric Constraints.

Geometric constraints control the relationship of objects with respect to each other. For example, you could assign the parallel constraint to 2 individual lines to assure that they would always remain parallel.

The following is a list of Geometric Constraints available and their icons:

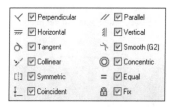

You can apply geometric constraints to 2D geometric objects only. Objects cannot be constrained between model space and paper space.

When you apply a constraint, two things happen:

1. The object that you select adjusts automatically to conform to the specified constraint.
2. A gray constraint icon displays near the constrained object.

Example:

If you apply the **Parallel** constraint to lines **A** and **B** you can insure that line **B** will always be parallel to line **A**.

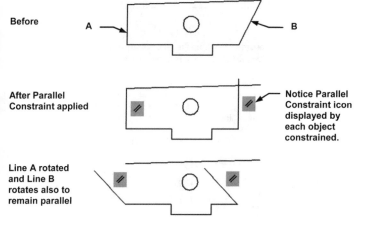

GEOMETRIC CONSTRAINTS....continued

HOW TO APPLY GEOMETRIC CONSTRAINTS

The following is an example of how to apply the Geometric Constraint **Parallel** to 2D objects.

The remaining Geometric Constraints are described on the following pages.

1. Draw the objects.
   Geometric constraints must be applied to **existing** geometric objects.

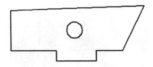

2. Select the **Parametric** tab.

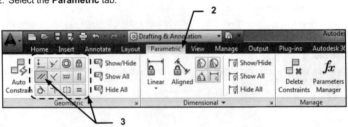

3. Select a Geometric constraint tool from the Geometric panel.
   *(In this example the **Parallel** constraint has been selected.)*

4. Select the objects that you want to constrain.
   In most cases the order in which you select two objects is important. Normally the second object you select adjusts to the first object. In the example shown below **Line A** is selected first and **Line B** is selected second. As a result **Line B** adjusts to **Line A**.

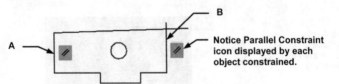

Notice Parallel Constraint icon displayed by each object constrained.

GEOMETRIC CONSTRAINTS....continued

COINCIDENT CONSTRAINT

A coincident constraint forces two points to coincide.

1. Draw the objects.

2. Select the **Coincident** tool from the Geometric panel.

3. Select the first **Point**.
   (Remember it is important to select the points in the correct order. The first point will
   be the base location for the second point.)

First Point

4. Select the second **Point**.
   (Remember it is important to select the points in the correct order. The second point
   will move to the first point.)

Second Point

The points are now locked together. If you move one object the other object will move
also and the points will remain together.

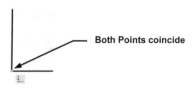

Both Points coincide

GEOMETRIC CONSTRAINTS....continued

COLLINEAR CONSTRAINT

A collinear constraint forces two lines to follow the same infinite line.

1. Draw the objects.

2. Select the **Collinear** tool from the Geometric panel.

3. Select the first line.

   (The first line selected will be the base and the second line will move.)

4. Select the second line.

   (The second line will move in line with the first line selected.)

The two lines are now locked in line. If you move one line the other line will move also and they will remain **collinear**.

GEOMETRIC CONSTRAINTS....continued

CONCENTRIC CONSTRAINT

A concentric constraint forces selected circles, arcs, or ellipses to maintain the same center point.

1. Draw the objects.

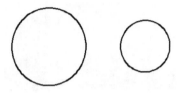

2. Select the **Concentric** tool from the Geometric panel.

3. Select the first circle.
   (The first circle selected will be the base and the second circle will move.)

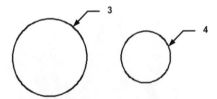

4. Select the second circle.

The second circle moves to have the same center point as the first object.

GEOMETRIC CONSTRAINTS....continued

FIXED CONSTRAINT

A Fixed constraint fixes a point or curve to a specified location and orientation relative to the Origin (World Coordinate System, WCS).

1. Draw the object.

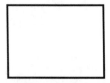

2. Select the **Fixed** tool from the Geometric panel.

3. Select the Fixed location.

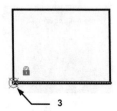

The bottom left corner is now fixed to the specified location but the other three corners can move. (**Note:** the "**Lock**" symbol may appear in a slightly different location)

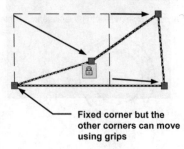

Fixed corner but the other corners can move using grips

GEOMETRIC CONSTRAINTS....continued

PERPENDICULAR CONSTRAINT

A Perpendicular constraint forces two lines or polyline segments to maintain a 90 degree angle to each other.

1. Draw the objects.

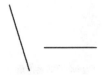

2. Select the **Perpendicular** tool from the Geometric panel.

3. Select the First object.
   (The first object will be the base angle and the second will rotate to become perpendicular to the first.)

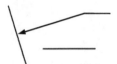

3. First object

4. Select the Second object.

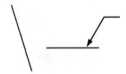

4. Second object

The second object rotated to become perpendicular to the first object.

GEOMETRIC CONSTRAINTS....continued

HORIZONTAL CONSTRAINT

The Horizontal constraint forces a line to remain **parallel to the X-axis** of the **current UCS**.

1. Draw the objects.

2. Select the **Horizontal** tool from the Geometric panel.

3. Select the **Object**.

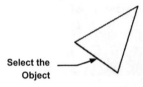

Select the _____
Object

The object selected becomes <u>horizontal to the X-Axis of the current UCS</u>.

Note: If the object is a Polyline or Polygon use the **"2 point"** option. The **first** point will be the **pivot** point.

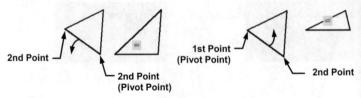

2nd Point

2nd Point
(Pivot Point)

1st Point
(Pivot Point)

2nd Point

GEOMETRIC CONSTRAINTS....continued

VERTICAL CONSTRAINT

The Vertical constraint forces a line to remain **parallel to the Y-axis** of the **current UCS**.

1. Draw the objects.

2. Select the Vertical tool from the Geometric panel.

3. Select the **Object**.

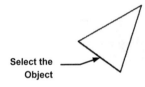

**Select the
Object**

The object selected becomes <u>vertical to the Y-Axis of the current UCS</u>.

Note: If the object is a Polyline or Polygon use the **"2 point"** option. The **first** point will define the **pivot** point.

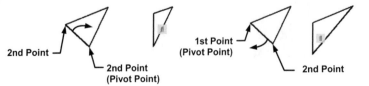

2nd Point

**2nd Point
(Pivot Point)**

**1st Point
(Pivot Point)**

2nd Point

GEOMETRIC CONSTRAINTS....continued

TANGENT CONSTRAINT

The Tangent constraint forces two objects to maintain **a point of tangency** to each other.

1. Draw the objects.

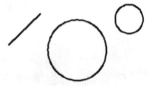

2. Select the **Tangent** tool from the Geometric panel.

3. Select the base object (Large Circle) and then the object (Line) to be tangent.

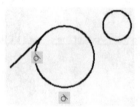

4. Select the base object (Large Circle) and then the object (Small Circle) to be tangent.

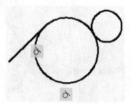

GEOMETRIC CONSTRAINTS....continued

SYMMETRIC CONSTRAINT

The Symmetric constraint forces two objects on a object to maintain **symmetry about a selected line**.

1. Draw the objects.

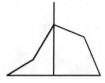

2. Select the **Symmetric** tool from the Geometric panel.

3. Select the base line (**1**) then the Line (**2**) to be symmetrical and then the line (**3**) to be symmetrical about.

4. Select the base line (**1**) then the Line (**2**) to be symmetrical and then the line (**3**) to be symmetrical about.

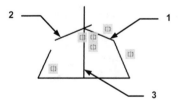

GEOMETRIC CONSTRAINTS....continued

EQUAL CONSTRAINT

The Equal constraint forces two objects to be **equal in size**. (Properties are not changed)

1. Draw the objects.

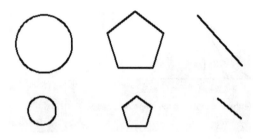

2. Select the **Equals** tool from the Geometric panel.

3. Select the base object (**A**) then select the object (**B**) to equal the selected base object.

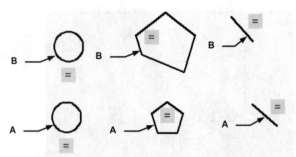

GEOMETRIC CONSTRAINTS....continued

CONTROLLING THE DISPLAY OF GEOMETRIC CONSTRAINT ICONS

You may temporarily Hide the Geometric constraints or you may show individually selected constraints using the **Show** and **Hide** tools.

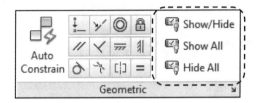

Show All

This tool displays all geometric constraints.
Click on the tool and the constraints appear.

Hide All

This tool hides all geometric constraints.
Click on the tool and the constraints disappear.

Show / Hide

After you have selected the **Hide All** tool to make the constraints disappear, you may display individually selected geometric constraints.

1. Select the **Show/Hide** tool.

2. Select the object.

3. Press **<enter>**

4. Press **<enter>**

The geometric constraints for only the selected objects will appear.

DIMENSIONAL CONSTRAINTS

OVERVIEW OF DIMENSIONAL CONSTRAINTS

Dimensional constraints determine the distances or angles between objects, points on objects, or the size of objects.

Dimensional constraints include both a name and a value.

Dynamic constraints have the following characteristics:
- A. Remain the same size when zooming in or out.
- B. Can easily be turned on or off.
- C. Display using a fixed dimension style.
- D. Provide limited grip capabilities.
- E. Do not display on a plot.

There are 7 types of dimensional constraints. (They are similar to dimensions)
1. Linear
2. Aligned
3. Horizontal
4. Vertical
5. Angular
6. Radial
7. Diameter

The following is an example of a drawing with dimensional constraints. The following pages will show you how to add dimensional constraints and how to edit them.

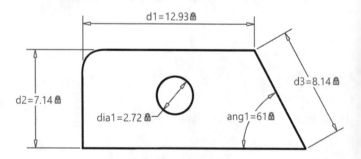

DIMENSIONAL CONSTRAINTS....continued

HOW TO APPLY DIMENSIONAL CONSTRAINTS

The following is an example of how to apply the dimensional constraint **Linear** to a 2D object.

1. Draw the objects.

2. Select the **Parametric** tab.

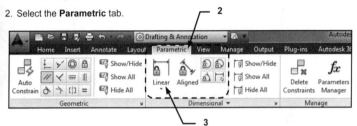

3. Select a Dimensional Constraint tool from the Dimensional panel.
   (In this example the **Linear** tool has been selected)

4. Apply the **"Linear"** dimensional constraint as you would place a linear dimension.
   a. Place the first point.
   b. Place the second point.
   c. Place the dimension line location.

5. Enter the desired value or **<enter>** to accept the displayed value.
   (Notice the constraint is highlighted until you entered the value)

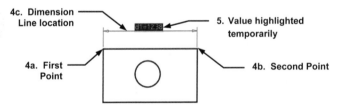

6. The Dimensional constraint is then displayed with **Name (d1)** and **Value (12.38)**.

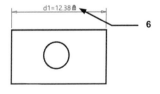

DIMENSIONAL CONSTRAINTS....continued

HOW TO USE DIMENSIONAL CONSTRAINTS

The following is an example of how to use dimensional constraint to change the
dimensions of a 2D object.

1. Draw the objects.
   (Size is not important. The size will be changed using the dimensional constraints.)

2. Apply Dimensional constraints.

   Linear and **Diameter**.

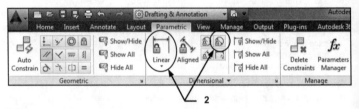

Note:
Dimensional constraints are very faint and will not print. If you want them to print refer
to page 13-24

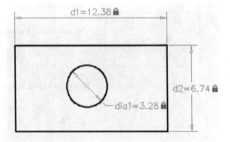

Continued on the next page...

DIMENSIONAL CONSTRAINTS....continued

Now adjust the length and width using the dimensional constraints.

1. Double click on the **d1 dimensional constraint**.

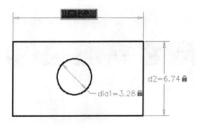

2. Enter the new value for the length then press **<enter>**

Note: The length increased automatically in the direction of the **"2nd endpoint"**. If you want the length to change in the other direction you must apply the **geometric constraint "Fixed"** to the right hand corners. (See page 13-8)

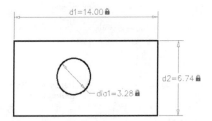

3. Double click on the **d2 dimensional constraint** and enter the new value for the width then press **<enter>**

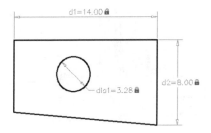

DIMENSIONAL CONSTRAINTS....continued

PARAMETER MANAGER

The Parameter Manager enables you to manage dimensional parameters. You can change the name, assign a numeric value or add a formula as its expression.

1. Select the **Parameter Manager** from the **Parametric tab / Manage Panel**.

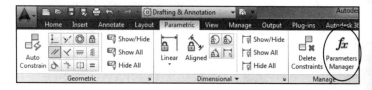

COLUMN DESCRIPTIONS:

Name: Lists all of the dimensional constraints. The order can be changed to ascending or descending by clicking on the up or down arrow.

Expression: Displays the numeric value or formula for the dimension.

Value: Displays the current numeric value.

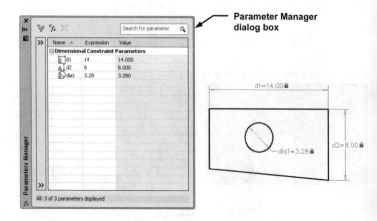

Parameter Manager dialog box

DIMENSIONAL CONSTRAINTS....continued

PARAMETER MANAGER NAME CELL

You may change the name of the dimension to something more meaningful. For example you might change the name to **Length** and **Width** rather than **d1** and **d2**.

1. Double click in the **Name Cell** that you want to change.

2. Type the new name and press **<enter>**

Example:

The names in the Parameter Manager shown below have been changed. Notice the dimensional constraints in the drawing changed also.

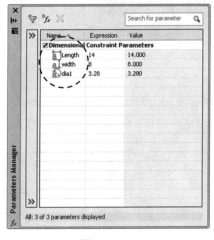

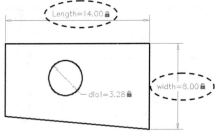

DIMENSIONAL CONSTRAINTS....continued

PARAMETER MANAGER EXPRESSION CELL

You may change the value of a dimensional constraint by clicking on the Expression cell and entering a new value or formula.

1. Open the **Parameter Manager**.

2. Double click in the **Expression Cell** to change (**dia1** in this example)

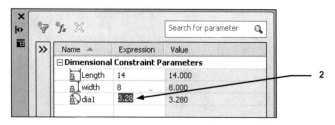

2

3. Enter new value or formula.
   For this example: **width/2**
   This means the diameter will always be half the value of **d2** (width).

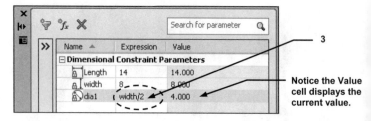

3

Notice the Value cell displays the current value.

Now whenever the width is changed the diameter will adjust also.

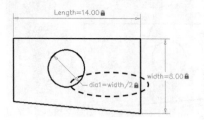

DIMENSIONAL CONSTRAINTS....continued

ADDING USER-DEFINED PARAMETERS

You may create and manage parameters that you define.

1. Open the **Parameter Manager**.

2. Select the **"New user parameter"** button.
   A **"User Parameters"** will appear.

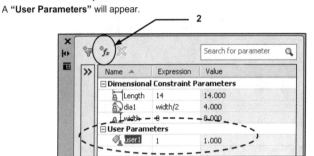

3. Enter a desired **Name** for the expression.

4. Enter an **Expression**.

The <u>Value cell</u> updates to display the current value.

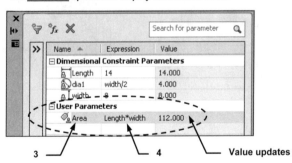

Note: With Imperial Units, the parameter manager interprets a minus or a dash (-) as a unit separator rather than a subtraction operation. To specify subtraction, include at least one space before or after the minus sign.

For example: To subtract **9"** from **5'**, enter **5' -9"** rather than **5'-9"**

DIMENSIONAL CONSTRAINTS....continued

CONVERT A DIMENSIONAL CONSTRAINT TO AN ANNOTATIONAL CONSTRAINT

Geometric and Dimensional constraints **do not plot**. If you would like to plot them you must convert them to an **Annotational Constraint**.

1. Select the **Constraint** to convert. (Click on it once)

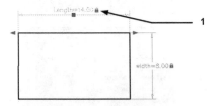

2. Right click and select **Properties** from the list.

3. Select the **Constraint Form** down arrow and select **Annotational**.

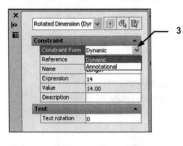

4. The Properties palette is populated with additional properties as the constraint is now an Annotational Constraint.

5. The Annotational Constraint **will now plot**.

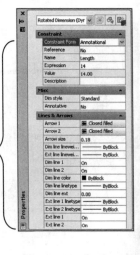

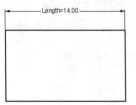

DIMENSIONAL CONSTRAINTS....continued

CONTROL THE DISPLAY OF DIMENSIONAL CONSTRAINTS

You may turn off the display of Dimensional constraints using the **Show** and **Hide** tools.
(Refer to page 13-15).

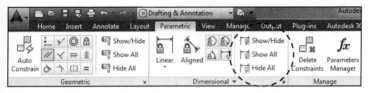

DELETE A DIMENSIONAL CONSTRAINT

To permanently delete a dimensional constraint select the **Delete Constraints** button
and then select the constraint to delete.

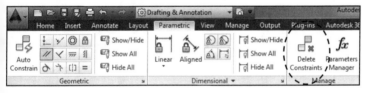

NOTES:

INDEX

A

Absolute Coordinates, 7-3
Adaptive Suggestions, 8-19
Add a Printer/Plotter, 5-2
Adjusting Viewport scales, 2-7
Aligned dimensioning, 3-23
Alternate Units, 3-49
Angular dimensioning, 3-24
Annotative Hatch, 10-18
Annotative Objects, 10-7
Annotative Property, 10-6
Annotative scale, Remove, 10-16
Annotative, Multiple scales, 10-12
Arc Lengths, Dimension, 3-25
Arc, 4-2
Array, 1-2
Array, Path, 1-11
Array, Polar, 1-7
Array, Rectangular, 1-3
Arrow, Flip, 3-32
Associative Dimensions, 3-2
AutoComplete 2014, 8-18
AutoCorrect 2014, 8-17
AutoCorrect List, Edit 2014, 8-17
Automatic save, 5-23

B

Back up files, 5-24
Back up files, 8-3
Background Mask, 8-2
Background Plotting, 9-2
Baseline dimensioning, 3-21
Block, Re-defining, 4-15
Blocks, 4-9
Blocks, Inserting, 4-13
Break, 1-13
Breaks, Dimension, 3-35

C

Centermark, 4-22
Chamfer, 1-14
Circle, 4-23
Collect Multileader, 4-21
Columns, 11-10
Command Line 2014, 8-17
Constraints, 13-2
Constraints, Dimensional, 13-16
Constraints, Geometric, 13-3
Continue dimensioning, 3-22
Coordinate Input, 7-2
Coordinate, Absolute, 7-3
Coordinate, Relative, 7-4
Coordinates, Polar, 7-7
Copy, 1-16
Copy, Array option, 1-18
Copy using drag, 1-19
Customize Quick Properties, 8-16
Customize your mouse, 5-31

D

Datum Feature Symbol, 3-54
Datum Triangle, 3-55
DDE, 7-6
Deviation, dimension, 3-51
Diameter, Dimension, 3-28
Dim, Geometric tolerance, 3-52
Dim., Alternate Units, 3-49
Dim., Datum Triangle, 3-55
Dimension Arc Lengths, 3-25
Dimension Breaks, 3-35
Dimension Diameter, 3-28
Dimension Large Curve, 3-27
Dimension line, Jog a, 3-38
Dimension position, Edit, 3-15
Dimension Radii, 3-30
Dimension Style, Create, 3-5

Dimension Style, Edit, 3-17
Dimension Style, Override, 3-18
Dimension Styles, 3-4
Dimension Sub-Style, 3-12
Dimension text, Edit, 3-14
Dimension, Aligned, 3-23
Dimension, Angular, 3-24
Dimension, Baseline, 3-21
Dimension, Continue, 3-22
Dimension, Deviation, 3-51
Dimension, Editing, 3-16
Dimension, Limits, 3-51
Dimension, Linear, 3-20
Dimension, Quick, 3-33
Dimension, symmetrical, 3-51
Dimensional Constraints, 13-16
Dimensioning, Ordinate, 3-45
Dims, Associative, 3-2
Dims, distance between, 3-39
Dims, Exploded, 3-2
Dims, Non-Associative, 3-2
Dims, Re-Associate, 3-3
Dims, Tolerances, 3-50
Direct distance entry, 7-6
Display UCS icon, 12-2
Distance between dims, 3-39
Divide, 1-20
Donut, 4-25
Donut, Fill mode, 4-25
Drag, copy using, 1-19
Drag, move using, 1-32
Drawing Limits, 10-2
Drawing Setup, 10-2
Dynamic Input, 7-11

E

Edit AutoCorrect List 2014, 8-17
Ellipse, 4-26
Erase, 1-21
Exit AutoCAD, 5-30
Explode, 1-22
Exploded Dims, 3-2
Extend, 1-23

F

File, Recover a, 5-24
Files, Back up, 5-24
Fill mode, 4-25
Fillet, 1-24
Flip Arrow, 3-32
Frames, Wipeout, 1-41

G

GDT text, 3-56
Geometric Constraints, 13-3
Geometric Symbols, 3-56
Geometric Tol. and Qleader, 3-53
Geometric tolerance, 3-52
Gradient Fill, 4-38
Grips, 8-4

H

Hatch, 4-29
Hatch, Annotative, 10-18
Hatch, Editing, 4-39
Hatch, Ignore objects, 3-44

I

Indents, Text, 11-9
Internet Search, Help 2014, 8-19
Input, Absolute, 7-3
Input, Coordinate, 7-2
Input, Direct distance, 7-6
Input, Dynamic, 7-11
Input, Polar, 7-7
Input, Relative, 7-4

J

Jog a dimension line, 3-38

L

Large Curve, Dimension, 3-27
Layer Match, 6-11
Layer, Color, 6-5
Layer, Lineweights, 6-6
Layer, Load Linetype, 6-9
Layer, New, 6-8
Layers, 6-2
Layers, Controlling, 6-3
Layout and Model, 2-2
Layout, Using a, 5-13
Limits, dimension, 3-51
Limits, Drawing, 10-2
Line spacing, 11-12
Linear dimensioning, 3-20
Lines, Drawing, 4-40
Lineweights, Layer, 6-6
Load Layer Linetype, 6-9
Lock a viewport, 5-12

M

Mask, Background, 8-2
Match Properties, 1-26
Match, Layer, 6-11
Measure, 1-27
Measure tools, 1-28
Mirror, 1-29
Mirrtext, 1-30
Model and Layout, 2-2
Mouse, Customize your, 5-31
Move using Drag, 1-32
Move, 1-31
Moving Origin, 12-3
Multileader and Blocks, 4-17
Multileader Style, 3-42
Multileader, 3-40
Multileader, add leader, 3-40
Multileader, Collect, 4-21
Multiline Text, 11-7
Multiline Text, Editing, 11-13
Multiple scales, Annotative, 10-12
Mutlileader, Align, 3-41

N

New drawing, Start a, 5-25
Non-Associative Dims, 3-2
Nudge, 1-33

O

Object Snap, 8-6
Object snap, Running, 8-10
Objects, Selecting, 5-32
Offset, 1-34
Open existing drawing, 5-18
Open multiple files, 5-19
Ordinate dimensioning, 3-45
Origin, moving, 12-3
Override Dimension Style, 3-18

P

Page Setup, 5-6
Pan, 8-11
Paragraphs, text, 11-12
Parametric drawing, 13-2
Path, Array, 1-11
Plotting from Model Space, 9-3
Plotting from Paper Space, 9-7
Plotting, Background, 9-2
Point, 4-43
Point Style, 1-20
Point Style, 4-43
Polar Array, 1-7
Polar coordinates, 7-7
Polar Snap, 7-10
Polar tracking, 7-9
Polygon, 4-44
Polyline, 4-45
Polyline, Editing, 4-49
Precision, Units and, 10-5
Printer/Plotter, Add a, 5-2
Properties palette, 8-13
Properties palette, Quick, 8-15
Properties, Match, 1-26
Purge, unused blocks, 4-16

Q

Quick Dimension, 3-33
Quick Properties Palette, 8-14

R

Radius, Dimension, 3-30
Re-Associate dims, 3-3
Recover a file, 5-24
Rectangular Array, 1-3
Rectangle, 4-51
Redo and Undo, 1-40
Relative coordinates, 7-4
Revision Cloud Style, 4-56
Revision Cloud, 4-54
Rotate, 1-36
Running object snap, 8-10

S

Save a drawing file, 5-22
Save, Automatic, 5-23
Scale, 1-37
Scaled drawings, create, 2-6
Selecting objects, 5-32
Setup, Drawing, 10-2
Single Line Text Editing, 11-17
Single Line Text, 11-14
Snap, Object, 8-6
Snap, Polar, 7-10
Snap, Running object, 8-10
Special Text Characters, 11-18
Spelling Checker, Text, 11-9
Stretch, 1-38
Symmetrical, dimension, 3-51
Synonym Suggestions, 8-19

T

Tabs, Text, 11-9
Template, Create a, 5-15
Template, Using a, 5-17
Text Style, Change effects, 11-6

Text Style, Delete a, 11-5
Text Style, New, 11-2
Text Style, Select a, 11-4
Text, Columns, 11-10
Text, GDT, 3-56
Text, Indents, 11-9
Text, Line spacing, 11-12
Text, Mirror, 1-30
Text, Multiline, 11-7
Text, Paragraph, 11-12
Text, Single Line Editing, 11-17
Text, Single Line, 11-14
Text, Special Characters, 11-18
Text, Spelling Checker, 11-9
Text, tabs, 11-9
Tolerances, 3-50
Tracking, Polar, 7-9
Trim, 1-39

U

UCS Icon, displaying, 12-2
UCS, moving the, 12-3
Undo and Redo, 1-40
Units and Precision, 10-5
Units, Alternate, 3-49

V

Viewport scales, 2-7
Viewport, Lock a, 5-12
Viewports, 5-9

W

Wipeout Frames, On or Off, 1-41
Wipeout, 1-41

Z

Zoom, 1-42